# The Newbie's Guide To Cannabis & The Industry

CHRIS CONRAD & JEREMY DAW

reset.me

*The Newbie's Guide to Cannabis & The Industry*

reset.me



3101 Clairmont Road • Suite G • Atlanta, GA 30329



Correspondence concerning this book may be directed to Reset.me, Attn: The Newbie's Guide to Cannabis & The Industry, at the address above.

ISBN: 0794843751
Printed in China

Please visit our website at Reset.me.

# CONTENTS

# WELCOME TO THE WORLD OF CANNABIS

Welcome to the world of cannabis, newbies. If you are not a newbie, welcome back. If you have friends and family members who use cannabis, this book will give you plenty to talk about with them. And if you are interested in trying cannabis or getting involved in this emerging industry, this book is meant for you no matter your level of interest, from curiosity to entry level to experienced to excelsior.

Every year, millions of people are discovering new ways to build a positive relationship with cannabis, including many people you already know. The story of cannabis and how hemp (arguably the world's most useful plant and planet-friendly farm crop) became the target of the long, failed drug war is a sordid tale of deception that we will tell you about later.

We're going to be telling you all sorts of important stuff as we go along. The facts might not match your expectations — and they certainly are not as lurid or as glamorous as they have been made out to be for these many decades of "Reefer Madness."

People are becoming more familiar with cannabis in society. Even as we type these words, the authors are aware of the paradoxes facing cannabis in its partially legalized status, and yet we are convinced that cannabis is here to stay. Four states have already voted to legalize adult use, and we hope this trend continues.

We want you to be comfortable with the basic ins and outs of cannabis, and with regular people who just like to have cannabis in their lives. This is an essentially benevolent plant and, likewise, there are really nice people who consume cannabis and are responsible, voting, taxpaying and contributing members of society. We will try to answer your questions and feed you some of the questions you may need to ask as we rise into the legalized world.

Marijuana products are not for kids; they are intended for consumption by adults only, unless a doctor recommends cannabis for medical use.

We see several distinct groups within the cannabis community: Casual consumers, connoisseurs, patients, medical marijuana interests, the

## Glossary

Marijuana is a word commonly used to sensationalize the female flowers of some cannabis strains.

Being "high" means that cannabis has been consumed and is causing a euphoric psychotropic effect.

Cannabis, Hemp, Marijuana; a plant with many names: bud, cheeba, chronic, dope, ganja, grass, herb, mota, mary jane, muggles, nugs, pot, reefer, skunk, the kind, weed, etc.

Byproducts include shake, dabs, edibles, medibles, extract, FECO, hash, wax, oil, BHO, RSO, phoenix tears, honeycomb, shatter, peanut butter.

There is a full glossary in the back of the book.

legalization movement, the cannabis industry and the hemp industry. Some are entrepreneurs and a growing number of people have jobs in the emerging cannabis industry. And, in every case, the cannabis community's interests are ultimately harmonious with those of society in general. We all agree that marijuana is not for kids unless there is a medical issue, so when we talk about marijuana, we are talking about adult usage.

**The first thing every cannabis newbie needs to know ...**

... Is that there will always be more to know. This book will take you through the basics and provide insightful details of useful information, but no matter what we tell you, there is a lot more information on any given topic we will address. We're not giving advice on legal issues or stock offerings; we're just presenting the options. As you move beyond the depth of information we present, you will need to seek it out.

It's a wide world of cannabis out there. Let's go exploring!

The advice in this book is not intended to get you into any trouble. While some policies are changing, marijuana is still illegal under federal law, even for medical use. Federal penalties include mandatory minimum sentences (MMS) of five years for 100 plants or 100 kilos, and 10 years for 1,000 plants or 1,000 kilos, and for stipulated dry weights, up to multiple life sentences if any guns are around. Proceed with caution and check out *The Citizen's Guide to State-by-State Marijuana Laws* at Whitman.com.

# 1 MEET THE CANNABIS PLANT

Cannabis hemp is an attractive, seed-bearing herb that is related to hops. It is the only plant in its genus and has but one recognized species — with hundreds of varieties and thousands of commercial uses. For our purposes, hemp or industrial hemp refers to its non-drug strains and products; marijuana refers to its various psychotropic and medical uses, and cannabis is the scientific name that covers all of its uses.

If you are a trained botanist or have a lifetime of experience farming other plant species — or even if you're a doctor who has personally observed the effects of other medications for decades — the qualities of cannabis hemp may very well astound you. In most ways, it is exceptional; in some, it is wholly unique. Join us now as we contemplate this modest plant and see what it has to offer us.

**SCIENCE FOR NEWBIES**

Tetrahydrocannabinol, THC, is the main psychotropic compound in cannabis.

Most Americans have never even seen a cannabis plant, although, were it not for a tragic accident (some say conspiracy) of history, hemp would literally be growing all around us in huge fields and nestled in gardens and flowerpots, cleaning the environment and powering whole swaths of the U.S. economy today. Instead of importing nearly half of our fuel from oil fields in the politically volatile Middle East, we could be growing hemp as a renewable energy crop in our own back yards — a plan proposed by American business icon Henry Ford himself. Instead of relying on potent but dangerously addictive opiate-based pharmaceutical painkillers for our injured veterans of foreign wars, we could be giving them cannabis — which is nontoxic and carries a risk of dependence roughly equivalent to that of coffee. Instead of having to choose between preserving our precious forests and providing homes where Americans can raise their families, we could be producing sturdy and environmentally friendly houses out of hemp (as is already being done, with great success, in Europe). Instead of offering only alcohol-serving bars and taverns as social gathering places to ingest psychotropic drugs, responsible adults would have the freedom to choose cannabis as a safer alternative. All of these benefits are just now coming back within our reach as a society, and this barely scratches the surface of what is possible with cannabis. So let's meet the plant again, like it's for the first time.

## EVERY NEWBIE NEEDS TO KNOW:

Cannabis is one of the fastest-growing plants on Earth. It can grow up to 20 feet in one season.

Cannabis is far and away the most useful plant known to humankind. *Popular Mechanics* counted over 25,000 commercial uses for the plant — in 1938! The list has expanded significantly since.

The human body has a specialized network of receptors in the brain and throughout the body that rely on compounds that match up with the cannabis plant's phytocannabinoids. These CB receptor sites and related compounds comprise the human endocannabinoid system, which was first discovered in the 1990s.

## ANATOMY & PHYSIOLOGY

Cannabis is an herbaceous annual plant that is dioecious, which means that it produces male plants separately from the female plants, and sometimes a single plant, called a hermaphrodite, may have both male and female flowers. The male produces pollen and the female does the rest of the work — producing fiber, resin and seeds. Its fruit is a seed kernel tucked into a leafy pouch that is coated with sticky resin glands. Let's start with a look at its component parts.

Botanical drawing of cannabis depicts the male plant (A), the female plant (B), a close-up of the male flower (1-3), the developing female flower (5-7, descending), male pollen spores (4), and both the development (11-13) and emergence (8-10) of a new seed. Credit: W. Müller

1. Seeds: Cannabis seeds contain the genetic material that determines the plant's characteristics, but they do not contain any psychoactive compounds. The seeds can be used in foods or pressed for oil, as discussed on the chapter on industrial hemp. Once upon a time in the U.S., almost every bag of marijuana purchased contained seedy bud from Mexico but, with the rise of sinsemilla bud, seeds became more scarce.
2. Roots: When grown from seed (figure 9, 10 opposite page), the plant puts out a taproot and a fibrous root ball. The taproot digs deeply to bring up nutrients and moisture from deep in the soil, while the root ball helps stabilize the plant, provide water and nutrients, and aerate the soil. When grown from a cutting, clones have no taproot, just the root ball.
3. Stalk: The plant's stalks are lobed, with a woody cortex wrapped in a bark containing long bast fibers. The cortex, or hurds, is composed of about 40 percent cellulose, while the bast fibers contain well over 60 percent cellulose and form a rough bark at the base of the plant. Cannabis has three kinds of stalk: trunk stalk grown

Ripe, mottled seeds. Credit: Ravenhurst (Wiki)

Cannabis grown from seed. Credit: Chris Conrad

from the base, branch stalks that divide into thinner and thinner branches as they reach away from the trunk, and petiole stalk that attaches all the leaves and flowers to the trunk or branches.

4. Leaves: When a cannabis seed sprouts, it puts out a taproot, a stalk and a set of round, dicotyledon leaves. The cotyledons send out an opposing pair of single lanceolate leaf blades with serrated edges and a mid-vein attached to petiole stalks; after these come another set of leaves: triple lanceolate blades attached to the stalk by petioles.

   The next set of leaves emerges at a 90-degree angle to the previous one and has five blades. The iconic cannabis leaves are compound and palmated, with an odd number of serrated leaflets, that we will call "fan leaves" for short. Early leaves emerge from the trunk in decussate-opposite pairs, meaning one on each side of the stem. It has a vertical stalk rising between the pairs. Pairs sprout from the stalk parallel to each other and at a right angle to the previous set; later on in the branching stage they begin to alternate from side to side as they ascend toward the top of the plant. Most of these fan leaves

Cannabis seedling grows its first (Fig A), second (Fig B) and third (Fig C) sets of true fan leaves. Credit: Chris Conrad

Prune the branch above the nodes to get more branching. Credit: Chris Conrad

typically range from five to 11 serrated blades per fan. These photos illustrate this process by looking down from above at the same plant at different stages of its vegetative growth. The leaves utilize and retain most of the fertilizer nutrients, so when they decompose they make rich compost.

5. Nodes: These small bumps of specialized cells occur at regular intervals on the trunk stalk and branches; under the right conditions, they can become leaves that offshoot either branches or flowers. If a cannabis plant's trunk is buried or if a branch end is cut to make cuttings for clones, the nodes can produce roots as well.
6. Branches: At every trunk or branch node, there is a potential set of new branches or flowers. Further along in the process, the male plants will produce stamens that release pollen while the females produce inflorescence, resin and buds — and even seeds, if they get pollinated.
7. Male flowers extend on petiole stalks from the plant's upper nodes opening into clusters of dusty yellow or greenish staminate flowers (figure 1 above) that shed copious amounts of pollen unless they are removed from the crop. These photos show the emerging male flowers appearing in the nodes and the spent mature plants that have already shed most of their pollen.

Male flowers almost ready to open. Time to remove it from the garden. Credit: Banana Patrol

This fully mature male plant has released millions of pollen spores. Fortunately, its growers kept it indoors. Credit: Chris Conrad

8. Female flowers grow in a pattern known as inflorescence, which refers to flowers clustered along a plant's upper branches. The pistillate female flower consists of a small green pouch called a calyx with two white stigmae hairs protruding (figure 5). As they mature, they cluster along the branch stalk and dendritic stems into dense forms that are known as bud, and the stigmae take on an amber color. The central bud at the very top of a female plant is considered to be its best and is called its cola, because someone once thought it resembles a fox tail. Nowadays, any relatively large bud can be called a cola.

Note the buds on this mature female plant. Credit: Chris Conrad

A female cannabis flower, just beginning to form. Credit: Chris Conrad

Close up of a glistening calyx. Note the resin glands, known as trichomes or hash. Credit: Banana Patrol

9. Resin is where the magic of cannabis happens. It is an excretion of terpenes and cannabinoids that is embedded in trace amounts throughout the plant (except its roots and seeds). The resin forms into tiny, nearly microscopic structures known as trichomes that look like spikes or mushrooms and is most concentrated in the mature female flowers. In fact, get a nice magnifying glass to look at them as they develop. The phytocannabinoids produced by the plants have therapeutic potential but THC-acid, from the live or raw plant, does not become psychotropic until it has been decarboxylated, usually by heating. So if you eat raw cannabis, it will not make you high, but if you

Female cannabis buds glisten with potent resin trichome. Credit: Chris Conrad

collect resin and decarboxylate it before using, there might be enough THC to have an effect (see Chapter Four's "Cannabinoid Chart"). The lower fan leaves of cannabis do not contain significant amounts of resin. There is a middle grade of leafy material and immature bud, called shake, from which lesser amounts of resin can be extracted; a higher grade of leafy material is called sugarleaf or trim. Regardless of how it is obtained, resin can then be extracted from the plant material to make one of many grades of cannabis concentrate.

Magnified many times, trichomes have a mushroom-like shape. Credit: Banana Patrol

## WHERE DOES HEMP FIT INTO THE EQUATION?

Hemp and marijuana have the same relationship as a Great Dane has to a Chihuahua — they come from the same species but have each been bred for such a uniquely different purpose that they seem totally different. And yet they can make babies together.

As with dogs, the domestication and genetic divergence between marijuana and hemp goes back a long time. Cannabis probably evolved somewhere around India (the origin of the word "indica"), and in these tropical climates it produced powerful medicine. But in the temperate latitudes of Europe the plant quickly adapted to different purposes. Under the more

Industrial strains of cannabis have long internodal stalks. Credit: Barbetorte

slanted sunlight, its therapeutic flowers reduced in size and potency, and its long fibrous stem elongated and strengthened. Before long, the Europeans adapted the lineage to a form better suited to their climate and needs. They called it "sativa" — meaning "useful" — but, in English, it took on another, simpler name: hemp. As opposed to medical strains (i.e., marijuana) that grow short, bushy and covered with resin in the tropics, hemp strains grow tall, straight and sturdy in the temperate climate zone. Medieval Europeans, and especially the mercantile Venetians of the Italian peninsula, mastered the art of transforming these long, sturdy fibers into rope strong enough to weather ocean storms and linens soft enough for a king. For many European centuries, the long and sturdy fibers of hemp's stalk have been its most prized product.

Farther north, a tough, scraggly, mold-resistant hemp strain known as cannabis ruderalis (roadside) grew wild and flowered on its own schedule. Still largely untamed, in our analogy, ruderalis might be compared to a street mongrel pooch that has potential but is not yet fully house trained.

Despite several efforts to reclassify cannabis genetics, most people still consider the plants to be of one species; but the differentiation of the strains and many sub-strains brings up the issue of cannabis mutations. We discuss this more in the breeding section.

## Glossary: Full glossary in the appendix

*Cannabinoids* are specialized terpenes found in the cannabis plant.

*Decarboxylate* means to remove a chemical chain from the plant molecule to metabolize it; most importantly for the reader, in the case of THC-acid, it must be decarboxylated to have a psychotropic effect.

*Decussate* leaves grow in opposite pairs, one on each side of the stalk.

*Flowers:* Cannabis has separate *pistillate* (female) and *staminate* (male) plants, and occasionally, a *hermaphrodite* plant has both sexes.

*Lanceolate:* a leaf shape that is narrow and tapered to a point.

*Palmate:* a leaf with multiple leaflets.

*Terpenes* are volatile plant compounds that produce fragrance or taste.

## Cannabis Misnomers

For most plants, the word "bud" refers to the unopened nascent flower, like a rosebud that opens into a blossom. For cannabis, the word bud refers to the mature and fully developed flower. Cannabis bud comprises mature flowers that form in dense clusters along the stalk nodes of the upper branches.

For all plants, the word "pollen" refers to powder released by the male (staminate) plants to pollinate the *stigmae* of the female (pistillate) plants and provide genetic material to produce seeds. However, people often use the word "pollen" to refer to kief, the powdery version of the collected resin glands of the female cannabis plant, which looks like ... well, pollen — but it's not.

Cannabis is non-toxic, but many people call it an "intoxicant." Likewise, when people think of the word "overdose," they often think of potentially lethal effects, but for cannabis, it just means that the experience has become *dysphoric*, meaning unpleasant; that the person has difficulty staying awake or, in more severe cases, that the individual may feel unwell, vomit up the edibles they just consumed or become so frightened that they call for medical help. You don't need a doctor to give you the prescription for recovery: drink some orange juice, eat some chocolate, get some rest, wait it out, and don't do that again!

# 2 HOW DID WE GET HERE?

To better understand where cannabis and the industry are going, let's look at how we got here.

## EVERY NEWBIE NEEDS TO KNOW:

"Cannabis has been a medicine for a lot longer than it hasn't been."
– *Dr. Donald Abrams, M.D.*

For thousands of years, people have used cannabis for food, clothing, housing, paper and industrial raw materials, according to archaeological evidence dating back about 10,000 years to the earliest known civilizations. Both in ancient China and Turkey, the bark fiber (bast) was used to reinforce pottery and bricks for housing. The fiber was likewise used to make cordage, textiles and caulking — the ropes, rigging, sails and sealants on seagoing vessels. Hemp was found to tolerate salt and moisture in the air, thus allowing sailors to stay at sea for longer periods of time. The Chinese were the first to write about the medicinal and culinary uses of *Da Ma* (true hemp) more than 5,000 years ago.

The USS *Constitution* earned its nickname "Old Ironsides" for its victory against the British ship *Guerriere* and is the longest-serving ship in the US Navy. It was outfitted with at least 60 tons of hemp rigging and sails.

The Hindu culture of India has been using *ganja* socially since about 1500 B.C. or earlier, mainly drinking it in sweet, milky beverages called

## The plant's name means "useful"

The Roman physician Galen used cannabis to treat the emperor. Dioscorides gave it the Latin name it bears to this day, cannabis sativa. "Cannabis" is its genus name and refers to the stalk; "sativa" means useful. It was adopted by Carl Linnaeus as Cannabis sativa, L, and is still used in our modern scientific system to identify and classify plants. The word "canvas" derives from the plant's name.

It was known as hanf to Germanic kings, hennep to Dutch merchants and hemp to English farmers and merchants, as it became the standard fiber of commerce. Other plants were generically labeled "hemp" to suggest that they had sufficient quality and tensile strength to be used for ropes, but only cannabis sativa could claim the title of "true hemp."

*bhang*. The Scythians and Thracians were pivotal in the history of cannabis, as noted by the Greek "Father of History," Herodotus, writing around 450 B.C. They grew hemp fiber and spun threads of such quality as to be confused with flax linens, he wrote, and held tribal purification rituals in which they went into huts billowing in cannabis smoke until they were cleansed and "emerged howling with laughter."

Our forebears were quite familiar with the sight and scent of cannabis growing in nearby fields or near the house to attract songbirds. The first paper was made with hemp, as were the sails that caught the wind to turn the windmills that ground wheat into flour and powered lumber mills. All that, and hemp was still just getting started.

Cannabis hemp has long been valued for its nutritional and non-psychoactive seeds containing essential proteins, essential fatty acids (EFAs) and edestin to help digestion. Recipes in many of the earliest known cookbooks discuss the use of hempseed as an ingredient, just as we find references to

> "How far... would there be propriety, do you conceive, in suggesting the policy of encouraging the growth of cotton and hemp in such parts of the United States as are adapted to the culture of these articles? The advantages which would result to this country from the produce of articles which ought to be manufactured at home is apparent."
>
> — *George Washington, October 1791 letter to Secretary of the Treasury Alexander Hamilton*

hemp in the first medical texts from China and Europe.

Many of the founders of the American Republic, including George Washington and Thomas Jefferson, grew hemp and knew its value. "Old Ironsides," the USS *Constitution*, carried 60 tons of hemp as sails (100 tons including rigging and caulk).

> **EVERY NEWBIE NEEDS TO KNOW:**
>
> "To know hemp in the U.S., one should understand its history, particularly legally. It is truly unique, tragic, ridiculous and surreal — all at the same time."
>
> *— Patrick Goggin, Industry Counsel and Vote Hemp Director*

Hemp kept America thriving for decades but eventually lost to the industrial revolution, when mass production of non-sustainable goods took over the economy. Instead of hempen sails powering ships and hemp cordage harnessing animals, smoky, coal-powered steam generators ravaged the earth in search of metals, gems, fossil fuels, timber and minerals. The plant lost market share and its production plummeted.

Meanwhile, ganja became a rising star of the medical world, and its tinctures and other products became staples in the 19th century doctor's chest after Dr. W.B. O'Shaughnessey's pioneering research on medical cannabis in India. Over the next 60 years, more than 100 reports on cannabis were published in scientific journals, and its use reached into the private chambers of Victoria, Queen of the British Empire. The difficulty of establishing reliable dosages and potencies was never resolved, and because the active compounds in cannabis are oil based, they were not suited for hypodermic injection by intravenous needles, as was then coming into vogue. Then, in a unique and unexpected moment of medical history, pharmaceutical drug companies moved to displace and ban plant therapies in the wake of the 1914 Harrison Narcotics Act.

As Henry Finger steered boards of pharmacy to act to take over medicine in the late 19th and early 20th centuries, Jim Crow laws were devised to target minority groups, petrochemicals moved to eliminate competing raw materials and prohibition took the nation by storm.

But hemp and cannabis medicines endured. At the U.S. Department of Agriculture, a fiber crop researcher named Lyster Dewey described in the 1916 *USDA Bulletin 404* his research showing four times greater efficiency in using hemp for papermaking than the use of trees. He called for a national

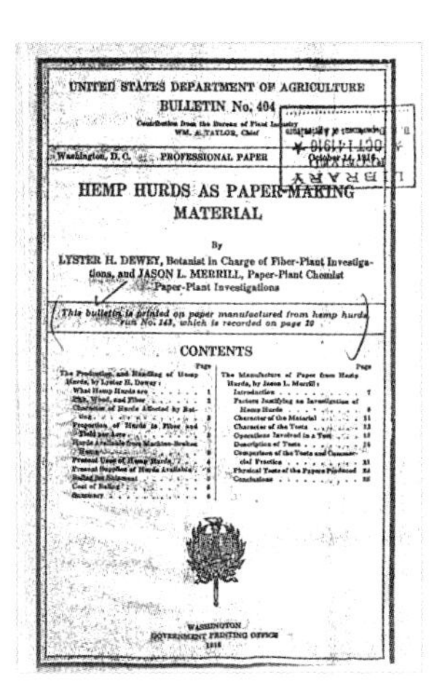

UNITED STATES DEPARTMENT OF AGRICULTURE
BULLETIN No. 404
Contribution from the Bureau of Plant Industry
WM. A. TAYLOR, Chief

Washington, D.C. PROFESSIONAL PAPER October 14, 1916

HEMP HURDS AS PAPER-MAKING MATERIAL

By

LYSTER H. DEWEY, Botanist in Charge of Fiber-Plant Investigations, and JASON L. MERRILL, Paper-Plant Chemist, Paper-Plant Investigations

(This bulletin is printed on paper manufactured from hemp hurds, run No. 143, which is recorded on page 20)

CONTENTS



WASHINGTON
GOVERNMENT PRINTING OFFICE
1916

The USDA knew of the benefits of making paper from hemp a century ago, when Lyster H. Dewey reported that hemp is four times as efficient as forestry for cellulose production for paper.

campaign of hemp farming to save both the family farms and the nation's forests. At the time, there was still only limited and inefficient farm machinery and factory equipment available to process hemp for the industrial age; but new inventions came into play in the Roaring Twenties that promised hemp would again become a major agricultural resource.

While this quiet revolution began to play out in America's farmlands, in the bustling cities the radical experiment of alcohol prohibition was failing spectacularly. In a few short years, it would come to an end, and a battalion of federal cops found themselves with no law to enforce. It looked like G-man jobs were on the chopping block halfway through the Great Depression.

During Prohibition, more people began smoking reefer at jazz clubs, but with the defeat of alcohol prohibition, the target was simply shifted to marijuana prohibition. The U.S. Department of the Treasury minions quietly brought a new tax law to Congress, the Marihuana Tax Act of 1937, at the behest of Harry J. Anslinger, director of the Federal Bureau of Narcotics, who orchestrated the media campaign that has become known as Reefer Madness. Anslinger tirelessly demonized the plant both before and after passage of the MTA, devising many of the propaganda devices ("what about the children") still in use today.

> "Cannabis indica does not produce dependence as in opium. ... There is no dependence or increased tolerance.... As with alcohol, it may be taken a relatively long time without social or emotional breakdown. Marijuana is habit forming, although not addicting in the same sense as alcohol might be with some people, or sugar or coffee."
>
> *— Dr. Walter L. Treadway, U.S. Public Health Service, Division of Mental Hygiene, 1937*

Familiar cannabis hemp became foreign-sounding "marihuana," and papers began to attach all manner of hysterical claims to the formerly mundane industrial workhorse. Anslinger and his goons prepared the MTA in secret over the course of two years and then snuck it through Congress despite outspoken opposition from the communities who learned too late that "marihuana" really meant cannabis hemp. The hemp industry and American Medical Association argued strenuously against the bill but lost.

For more than three decades, the MTA nominally required a special tax license to grow or transfer cannabis; except the only time the government actually issued licenses was during World War II, when the War Hemp Industries Program produced the film *Hemp for Victory* to encourage farmers to grow industrial hemp once again.

Over the course of time, the pharmaceutical drug industry consolidated its grip on medicine. The nation's forests were systematically cut by the logging industry; its rivers choked on foul-smelling sludge from pulp mills that digested forests into pulp for cheap paper. Burning coal and oil products contaminated the air and polluted the seas. The word "hemp" was being systematically erased from the global memory, and "marijuana" was inserted to take its place.

After WWII, Americans began to explore new parts of the globe, and some encountered new cannabis strains and discovered their psychotropic effect. Small amounts of potent bud were shipped in pickup trucks across the border from Mexico to jazz musicians and the Beat Generation, and later for hippies and college students. When Bob Dylan turned on the Beatles, a whole new era of cannabis use began.

The USDA produced *Hemp for Victory* in 1942, a film urging farmers to grow hemp for the national defense. Credit: U.S. Department of Agriculture

## Cannabis History Timeline, Part I

8000 B.C.: Prehistoric cannabis use: Hemp used as a fiber and food crop.

4000 B.C.: Ancient China: Hemp for foods, rope, textiles, medicines and paper.

1000 B.C.: Scythians, Thracians, Hindi, Zoroastrians and Coptics use cannabis for social repast and spiritual/religious sacrament.

1000 A.D.: Arabic and Mediterranean commerce and maritime use spread.

1450: Gutenberg uses hemp to make the first printed Bible; hemp paper books replace textile and animal skin parchments as a way to transport the written word.

1492, Age of Imperialism: Hempen sails, rigging and caulking are the standard for oceangoing vessels as Europe expands onto other continents to seize resources.

18th Century Age of Enlightenment: Books printed on hemp spread knowledge, provide the ledgers and substance of commerce; newspaper-type journals become available to common people, hempen sails power wind mills, hemp tents and uniforms are standards for explorers, hemp-covered wagons traverse the plains and deserts.

1791: The cotton gin makes cotton "king," and slavery increases fourfold.

1842: Dr. W.B. O'Shaughnessy rediscovers medical marijuana while working in India.

1914: International pharmacy boards move to ban herbal medicines, like cannabis.

1930s: Decorticator invented; hemp is again viable for agriculture.

1935: Anslinger launches "Reefer Madness" and "marihuana" demonized nationwide.

1937: Hemp banned under the name "marihuana."

1941: Prohibition suspended for licensed farmers during the War Hemp Industries Program for WWII.

1969: Marihuana Tax Act ruled unconstitutional.

The U.S. Supreme Court struck down the MTA in 1969 as an unconstitutional violation of the Fifth Amendment in *Timothy Leary v. USA*. Congress scrambled to replace the law with the Controlled Substances Act of 1970, which ostensibly set a scientific framework for a new federal law enforcement bureaucracy, the Drug Enforcement Administration (DEA), to rate all drugs by their relative safety and risks, and regulate them accordingly. Cannabis was placed "temporarily" in Schedule 1, for the most dangerous drugs only.

President Nixon launched a global War on Drugs in part to undermine the U.S. peace movement. Credit: White House

When the Shafer Commission came back in 1972 with its proposal that personal use and possession of up to an ounce of marijuana should be decriminalized, the DEA sat on its hands. The federal National Institute on Drug Abuse (NIDA) was created to make sure that only negative research on cannabis "abuse" would be allowed.

Nixon declared war on drugs and sealed off the border with Mexico, essentially handing control of the smuggling trade over to organized drug gangs. Keith Stroup formed the National Organization for the Reform of Marijuana Laws (NORML). Michael and Michelle Aldrich and Gordon Brownell formed the Amorphia group to import hemp cigarette papers that were sold to help fund California Proposition 19 to legalize marijuana, which was trounced at the ballot box. Across the Atlantic, the government of the Netherlands adopted a "tolerance" policy in which small cannabis retail purchases at coffee shops would not be prosecuted. Commerce flourished, and a good time was had by all.

Several U.S. states, on the other hand, have gone through an evolution of laws — starting in the 1970s when a flurry of reforms decriminalized small amounts of cannabis. Dr. Tod Mikuriya compiled the *Marijuana Medical Papers 1942 to 1972* and Dr. Lester Grinspoon wrote *Marihuana Reconsidered.* Tom Forcade founded *High Times* magazine to foster the soft drug culture in America. Cheech and Chong smoked giant joints on movie screens. Glaucoma patient Robert Randall got a medical necessity court order in 1976 that led to the creation of a national medical marijuana bureau, the Investigational New Drugs (IND) program; Randall went on to form the Alliance for Cannabis Therapeutics (ACT), but the scientific progress that group achieved was largely thwarted by NIDA obstructionism.

**"Penalties against possession of a drug should not be more damaging to an individual than the use of the drug itself."**
— *President Jimmy Carter, Aug. 2, 1977*

## Cannabis Timeline, Part II

1970s: NORML founded. Oregon is first state to decriminalize possession.

1978: President Jimmy Carter calls for federal decriminalization.

1988: Judge Young rules DEA has misclassified marijuana.

1992: San Francisco voters pass Proposition P, the first medical marijuana ordinance in the country.

1996: California voters pass Prop 215 to legalize medical use and cultivation. In 1998, five other states did the same.

2003: California SB 420 legalizes sales of marijuana among patient groups.

2009: Ten hemp states and 15 medical marijuana states.

2010: Californians narrowly defeat Prop 19, which would have legalized adult use and regulated sales.

2012: Colorado and Washington legalize tightly regulated sales for personal adult use. Only Colorado legalizes home grows.

2013: Uruguay announces that it will legalize marijuana.

2014: Medical use legal in 23 states but no federal relief. Oregon, Alaska and Washington, D.C., legalize adult use and some sales.

2015: Hemp is legal in 20 states, President Obama signs Farm Bill legalizing some hemp cultivation, Congress begins to cut funding for the DEA, Jamaica makes major reforms and Canada's new government announces it will legalize cannabis.

President Jimmy Carter proposed in 1978 that Congress should decriminalize an ounce of marijuana for personal use, but when Ronald Reagan took office, he brought back the drug war with a vengeance, the embodiment of big government at its worst. With the 1980s came crack cocaine, Zero Tolerance, Just Say No, urine testing laws to force cannabis consumers out of their jobs, the DARE program, Partnership for a Drug Free America, mandatory minimum prison sentences that tied the hands of the courts, conspiracy laws and property forfeiture. The ACT went on a public education campaign seeking to expand the IND program. With the help of NORML attorneys, ACT took a case to federal court to force the DEA to reschedule cannabis. They won, but the DEA refused to comply with the court's order. So, despite a subsequent 1988 court ruling that the agency was "arbitrary, cruel and capricious" to keep cannabis in Schedule 1, bureaucracy has prevailed. Subsequent presidents, courts and Congress have allowed the DEA to ignore science and assert its political authority to reclassify the plant, so the federal ban remains firmly in place.

George H.W. Bush pushed the Reagan agenda with an even heavier hand: aerial surveillance, thermal detectors, dogs, snitches. In a pique against hundreds of AIDS patients who were seeking to enter the IND program, Bush cancelled the program to all new patients, including those who had already been approved for medical necessity.

In the late 1980s, hemp struck back. Gatewood Galbraith campaigned as the perennial hemp candidate for governor in Kentucky. Ben Masel, author Jack Herer and the Cannabis Action Network (founded by Debby Goldsberry and Monica Pratt) began touring the country holding rallies calling for legalization. Oregon activists including Doug McVay and Paul Stanford attempted a combination hemp and marijuana initiative; they lost, but kept trying. The Drug Policy Foundation (founded by Kevin Zeese and Arnold Trebach) and the Lindesmith Center (headed by Ethan Nadelmann) began to work on a variety of policy issues. Dale Gieringer formed and became director of California NORML.

Also in the late '80s, Chris Conrad launched the Business Alliance for Commerce in Hemp (BACH) and began networking among the various stakeholders on a four-fold strategy of advancing industrial hemp, medical marijuana, legal adult cultivation and use and regulated adult cannabis commerce. By 1990 he had organized a network of environmentalists, farmers, students, professors, entrepreneurs and media representatives around the country, and sourced legal hemp products from abroad to encourage business development. He churned out photocopy-ready literature as BACH and the Family Council on Drug Awareness and collaborated on, designed and produced Herer's *The Emperor Wears No Clothes*, a book that reframed the legalization debate. Senator John Galiber of New York sponsored the first modern industrial hemp bill in a state legislature on behalf of the local BACH chapter. Steve Hager awakened *High Times* readers to the hemp issue.

In the early 1990s, two major organizations branched out from NORML: Patients Out of Time (POT), with Mary Lynn Mathre and Al Byrne, put its focus on veterans and medical research, going on to sponsor a biennial Clinical Conference on Cannabis Therapeutics. The Marijuana Policy Project, with Rob

**EVERY NEWBIE NEEDS TO KNOW:**

"Marijuana, in its natural form, is one of the safest therapeutically active substances known to man." — *DEA Administrative Law Judge Francis Young, 1988*

Kampia and Chuck Thomas, set its focus on elections and legislative sessions, playing a role in most of the successful state reform bills since then. Hempsters held thousands of hemp rallies in parks and in front of government buildings all over the country. Conrad and Herer launched the California Hemp Initiative to educate and mobilize the voter base and train grassroots petitioners. San Francisco passed Prop P to protect patients, and Dennis Peron opened the first approved medical marijuana dispensary, the S.F. Cannabis Buyers Club. Mike and Valerie Corral formed the Wo/Men's Alliance for Medical Marijuana (WAMM) to serve desperate patients in Santa Cruz, California. An elderly but feisty nurse, Brownie Mary Rathbun, got arrested delivering cannabis brownies to AIDS patients and fiercely defended her actions. Federal IND patients Elvy Musikka and Irvin Rosenfeld began speaking out on the benefits of medical use.

More than 40 recently formed hemp businesses founded the Hemp Industries Association (HIA) in 1994. Hemp became the darling of hippies and environmentalists. Hemp Clubs sprang up on college campuses around the country. Colorado became the first state to enact a hemp bill but could not get federal DEA licenses to import seeds or grow the crop. Farmers and industrialists with a political strategy of supporting hemp legalization while opposing marijuana legalization formed the North American Industrial Hemp Council. Hemp stores opened in towns all over the country; many lasted only a few years but left a lasting legacy of consumers looking for hemp products to buy.

Washington activists held their first Seattle Hempfest, which has since grown into an annual tradition and pilgrimage for much of the cannabis community. The years of hard work first paid off in 1996, when Peron and other Bay Area activists filed a medical-use initiative and Nadelmann convinced billionaire George Soros to join wealthy donors Peter Lewis, John Sperling and George Zimmer in funding the signature drive that put Proposition 215 on the California ballot. Volunteers collected nearly 200,000 of the required signatures. That November, voters for the first time legalized cultivation and possession of marijuana for medical use. President Bill Clinton admitted that he had once tried marijuana himself, but then went on to escalate arrests, prosecutions and drug war spending. He signed a law adding for the first time the federal death penalty for marijuana sales and cultivation — under which many of the nation's Founding Fathers would have been arrested and put to death for growing hemp. (Put that in your pipe and smoke it, King George III). When heavily armed police swept in on Peron's San Francisco club, the national media showed paramilitary officers in SWAT gear wielding guns and truncheons to round up sick people in wheelchairs.

Seattle Hempfest, pictured here, is the largest 'protestival' in the U.S.

At least four groups — Erowid.org, Dave Borden's DRCnet.org, Mark Greer's DrugSense.org/Media Awareness Project, and Cliff Schaffer's DrugWarLibrary.org — all launched campaigns to inundate the then-emerging Internet with reform activism, a move that paid back in droves. The reform movement began growing exponentially. Activists Mikki Norris, Chris Conrad and Virginia Resner launched Human Rights and the Drug War, a photo exhibit to show the faces and families of POWs and victims of the Drug War, which touched the conscience of the American people. In Oakland, California, Jeff Jones opened the Oakland Cannabis Buyers Co-op, and Richard Lee transplanted himself from Texas to the Oaksterdam District (Oakland + Amsterdam), known for its 12 dispensaries. New York yippie Dana Beale launched the Global Million Marijuana Marches, held on the first weekend of May every year in hundreds of locations around the world. The Deadheads and *High Times* popularized the number "420" as a code meaning that it's "time to get high" — and April 20 became a cannabis holiday. Actor Woody Harrelson and singer Willie Nelson began to champion hemp products as a

After the federal crackdown of 2010, a wave of protests swept federal buildings around the nation. The Department of Justice shifted a number of its policies over the next few years.

way to save the family farm. By the end of the decade there were scores of medical marijuana shops operating in the states. Even though marijuana was not legal yet, many people acted like it was and paid a heavy price. Growers and caregivers got decades-long prison sentences. Arrest rates continued to climb as most police officers found marijuana arrests to be a gateway to promotions and bonuses.

By 2000, California was joined by medical marijuana states Oregon, Washington, Alaska, Colorado, Hawaii, Maine and Nevada. Local organizations had spread to nearly every state, tearful patients were seen at rallies and testifying before state legislatures and, in 2002, Law Enforcement Against Prohibition (LEAP) brought retired judges and law enforcement agents to publicly speak against the Drug War. Don Duncan became a point person for dispensary activism, and Steph Sherer formed Americans for Safe Access to fight for patients' rights and access to cannabis. In 2003, California adopted SB 420, the first law to immunize patients growing and selling marijuana to each other in associations that came to be known as patient collectives, formalizing the storefront dispensary system for which the state is now famous.

Support from the U.S. Supreme Court since 2000 has been mixed. First the court recognized with the *Conant v McCaffrey* decision that doctors have a First Amendment freedom of speech right to recommend or approve medical marijuana to their patients. Then came the *Oakland Cannabis Buyers Cooperative* ruling, which held that there is no federal medical necessity defense for selling marijuana. Then came *Gonzalez v. Raich*, which held that the federal government can still arrest patients in states where medical use is legal. The decision rested on the dubious premise that, even if marijuana is grown in a state where it is legal with no exchange of money or crossing of state lines, federal intervention is enshrined in the interstate commerce clause because an imaginary person in a different state might think about buying it. In *HIA v DEA*, an Appeals court stopped the DEA attempt to ban hempseed food products and forced it to repay the hemp industries' legal costs. In *San Diego v California* the Supreme Court let stand a ruling that state criminal justice laws do not need to conform with federal law: in other words, there is no federal pre-emption over state drug laws.

So many people and events have joined the movement that it became impossible to keep up with them all. The DPF and Lindesmith Center united into the Drug Policy Alliance (DPA) under Nadelmann. Students for Sensible Drug Policy and NORML chapters activated college students to engage in overall drug policy reform. Colorado activist Mason Tvert spearheaded the SAFER campaign based on the notion that marijuana is safer than alcohol and should not be regulated more strictly. Steve DeAngelo took things to a new business level by helping launch Harborside Health Center, Steep Hill Labs and later the Arcview Group for investors. Expos focused less on the many uses of hemp and more on the business opportunities of medical marijuana.

Newly elected President Barack Obama promised to reduce federal pressure on medical marijuana states, and for one whole year in office, he held true to that promise. Ten states had industrial hemp bills by 2009; 15 had medical-use laws, as well as Washington, D.C., — the federal District of Columbia. For the first time, polls were showing that a plurality but not a majority of Californians backed legalization, and support for medical use soared in the 70 to 80 percent range. With a growing number of dispensaries opening in various medical-use states, Aaron Smith founded the National Cannabis Industries Association to lobby for reasonable regulations.

Up until this point, much of the progress for reform had been made by voter initiatives, though a growing number of bills had cleared state legislatures. The courts were little or no help at all. When the Montana legislature largely repealed the state's voter-enacted medical marijuana law in 2010, it triggered a massive federal crackdown. When Obama targeted medical distribution, public opinion polls shifted; the small plurality of support for legalization in national polls jumped to a small majority saying to legalize it and the let states decide.

California's Proposition 19 campaign came within 4 percent of legalizing adult use of cannabis in 2010 and changed the global debate.

Richard Lee founded Oaksterdam University in 2007 to educate people about the emerging cannabis industry and funded a new initiative in 2010 to legalize marijuana. The recent Prop 19 was narrowly defeated but succeeded in bringing mainstream media, the NAACP and union support behind marijuana reform. The campaign changed the national discourse on cannabis from *if* marijuana should be legalized to *how* it should be legalized. Building on the momentum, Steve Fox joined Tvert to lay the groundwork for a statewide initiative in Colorado, Amendment 64. A controversial and highly restrictive measure, I-502, made it onto the Washington state ballot with the support of Lewis, the ACLU and travel writer Rick Steves.

When voters in Colorado and Washington legalized adult possession and regulated sales in 2012, the federal crackdown slowed; but the growing pains of a new industry have not abated. Instead, investors generated a "green rush" while the two earliest legalizing states took divergent paths: Colorado generated massive tax revenues from cannabis, while Washington waffled and stumbled around for years setting up its program, then suppressed medical

In 2012, Georgia Edson (center) became a spokesperson as Colorado activists ran the first successful state initiative to re-legalize cannabis for adults, Amendment 64, including licensed sales. Credit: Yes on A-64

sales while it overtaxed and regulated adult distribution. Meanwhile, the high prices and lack of access allowed the underground market to continue as producers who were shut out of the legal market were willing to accommodate the national appetite for bud. Nonetheless, arrests and prosecutions for cannabis dropped.

Then, out of the blue, Uruguay's president announced that the small South American nation would legalize marijuana, as well. Unlike U.S. jurisdictions, which sought to suck as much money as possible from the cannabis market to keep prices inflated, Uruguay sought to bring the legal price down so low that the illicit market cannot compete with government-licensed suppliers.

Several years later, the world has not come to an end. Congress added an industrial hemp amendment to the Farm Bill signed by President Obama in early 2014. By that summer, 23 states allowed some legal medical use, and two more states, Alaska and Oregon, legalized cultivation, possession

**EVERY NEWBIE NEEDS TO KNOW:**

The genie is out of the bottle. And he's not going back in.
— *Paul Armentano, NORML deputy director*

and sales in November. A number of states began to enact CBD-only medical marijuana laws in an effort to cut off further legalization efforts.

Meanwhile, the DEA stumbled badly when it tried to block delivery of hemp seeds to Kentucky for a state-authorized program; the House of Representatives added hemp into the Farm Bill and another bill restricted the use of DEA funds to block state implementation of medical marijuana laws. When the 2015 budget was approved and signed into law, Congress included limits on the use of DEA funds against state-legal medical marijuana and industrial

## EVERY NEWBIE NEEDS TO KNOW:

"The cannabis entrepreneur must understand that, at this point, the entire industry rests on a legal house of cards. It is primarily the political strength of the reform movement that keeps the Justice Department from shutting down the cannabis industry in a short series of lightning actions. For every dollar invested in the business, at least a dime should be set aside to support political reform activities." — *Eric Sterling, Criminal Justice Policy Foundation*

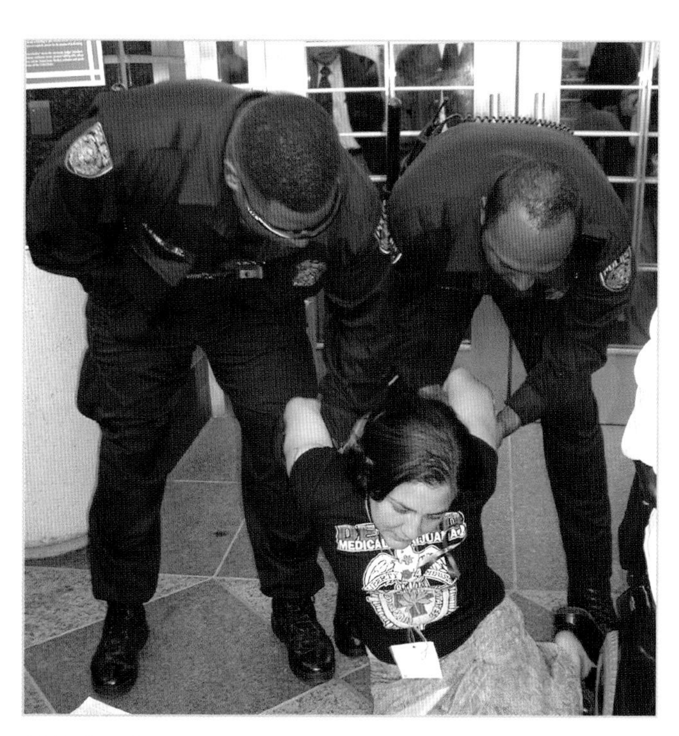

ASA's Step Sherer gets arrested for protesting federal raids on cannabis dispensaries in 2009 in Sacramento.

hemp programs. In April 2015, the vilified head of the DEA resigned, leaving an untested replacement—and thousands of Americans languishing in prison for years, decades or even life behind bars for growing and selling a plant.

That's when Jamaica announced that it would institute massive reforms for the medical, religious and personal use of cannabis (even for some tourists) and encourage hemp cultivation.

The current DEA director, Chuck Rosenberg, faces a very different horizon. There are fewer legalization rallies and more business conferences. Congress changed the definition of marijuana to allow hemp to be grown in states that allow it as long as the THC

David Bronner being arrested in Sacramento for protesting federal raids on cannabis dispensaries, 2009.

Inside the IndoExpo trade show, vendors showed the latest in equipment. This display shows the use of LED lamps instead of standard HIDs.

content is under 0.3 percent in the flowers. Likewise, new funding restrictions rein in some of the agency's most egregious anti-medical marijuana activities but are vague enough that no one is quite sure what they entail. The Obama administration has given a provisional green light to state-legal personal use laws in the four legal states, but Congress has not voted to enshrine that policy into law. In 2016, a new president will be elected; whether the current administration's policies will continue, expand or be terminated remains to be seen. Meanwhile, Canadian voters in 2015 swept the Liberal Party into power with a mandate to legalize marijuana north of the border.

This is the shaky foundation on which we find the hemp and cannabis industries working to build a secure future. If it all sounds a bit tenuous, that is because it is. But the genie is out of the bottle. Momentum is building around the world, and a new generation of mainstream Americans no longer buys into the lies and bigotry of marijuana prohibition. Likewise, every financial interest that can — big taxes, excessive packaging, complex security systems, etc. — seems to be eager to attach itself to cannabis.

Given how safe and innocuous cannabis really is, there is no reason for the excessive regulatory zeal of the times. We certainly hope that the façade of "protecting society" from this helpful plant will wither and die a quick death so that freedom, and cannabis, can grow.

A Denver freeway billboard advertises IndoExpo trade show events in 2014. Credit: Chris Conrad

# 3 CONSUMING CANNABIS

How do people consume cannabis?

Most people think of lips puckered up, sucking hard on a hand-rolled cigarette, eyelids half closed — then holding in the inhalation for as long as possible until it is released in a gasp for fresh air, followed by a sigh of relaxation and contentment. That is exactly what happens a lot of the time cannabis gets consumed, but there are many options available to newbies in the know.

There is smoking with or (preferably) without tobacco, vaporization and dabs; for ingestion, there's edibles, Full Extract Cannabis Oil (FECO) AKA phoenix tears, juicing and tinctures; and for topicals, there's liniment, sprays, creams and salves.

Whether your plan is to try cannabis, incorporate it into your life, or make it into a career, you will need to know the wide world of cannabis consumption. The dizzying diversity of cannabis pipes and accessories, from the most rustic to the most elaborate high-tech designs, is a lot for a newbie to take in.

First let's look at some of the ways people inhale cannabis.

## TOKING MARIJUANA

Cannabis smoking is called "toking" to differentiate it from tobacco, and this is an extremely effective method of bringing cannabinoids into your life.

Using either a pipe or rolling papers, the cannabis is loaded in and set aflame. Inhale deeply for up to five seconds, and exhale. The heat decarboxylates the THC-acid and converts it into psychotropic THC, which is carried from the smoking implement down the throat to the lungs where the cardiopulmonary system mixes it into the bloodstream and sends it throughout the body to the various cannabinoid receptors. The leftover smoke is exhaled from the lungs while the activated compounds go to work weaving a temporary shift of consciousness or its needed medical impact. The effect is nearly immediate and a person can stop after one, two or three inhalations (called "hits") — or they may puff away and smoke multiple times throughout the day, depending on tolerance, availability and preference or need. The THC converts into an inert metabolite and is safely stored in the body's fatty tissue before being discarded through the urine, feces or hair over a period of days or weeks. That's why it can be detected by a drug test long after the subjective effects have long gone, and why it doesn't show impairment, only whether someone consumed cannabis in the last few weeks or so.

Glass water pipes, or bongs, remain popular.

As a newbie, you have plenty of options. You could choose a joint (a hand-rolled cigarette with or without a filter), a pipe of often colorful glassware, or a water pipe, such as a hookah or a bong, which features a wide mouth, a water chamber and a ventilation hole called a carburetor. Some groups, and particularly Europeans, mix tobacco with cannabis before smoking (making a spliff), or wrap cannabis in a tobacco leaf (sometimes flavored), or hollow out a cigar and add marijuana to the contents, which is called a blunt. The upper right photo shows a bong, while the other images on the next page illustrate the relative sizes of some joints you may encounter. Notice the set-aside bits of stem that were removed from the bud before rolling the cigarettes to prevent tears in the paper.

A safe smoking technique is to hold the joint between cupped fingers, making a fist pipe, and inhale through the gap between forefinger and thumb, chillum style.

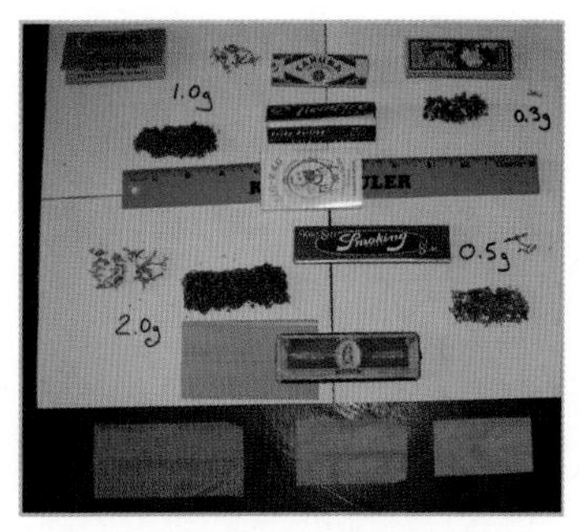


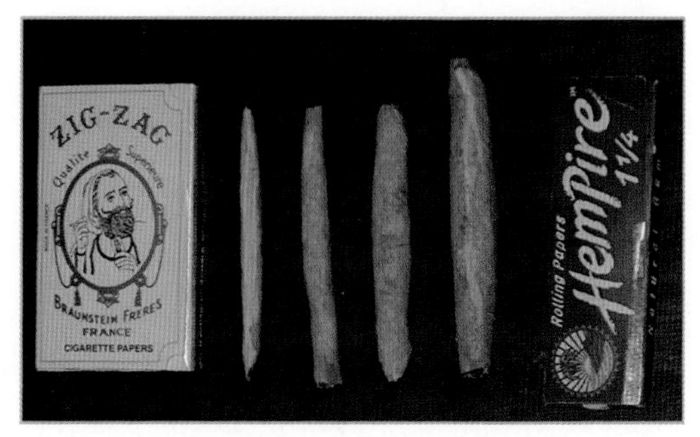


You might encounter handrolled joints in various sizes.

## STRAIN SELECTION

Potency can vary greatly from bud to bud, and a lower potency can often be more enjoyable than the stronger stuff. Different kinds of cannabis have different effects, so which is for you? A tried and true way to decide is by sampling strains. Retailers and dispensaries usually want to keep their clientele happy by having on hand a variety of strains and a knowledgeable "budtender" to help customers make that selection.

Ask for a suggestion, or you may want to pick out a selection of strains of bud or resin products, get a nice clean pipe out and break off a piece of one of the strains, keeping the other strains separated. Light up, take one or two inhalations and then stop to assess the effect. It's normal to cough at some point. If you feel suddenly lightheaded, that's a "rush." Take it easy, newbie, because maybe that's enough for one day. Another way to get a rush is to be sitting or squatting when you take a toke and then stand up quickly, causing transpostural hypotension. Time to sit back down.

Try the next strain and repeat. It is best to pause an hour or so between varieties, but most people don't wait that long. The problem is that if you try one sample right after the other, it is hard to tell which strain is affecting you; is the second taste actually stronger or is it compounding the effect from the first one? It's hard to say without a delay.

Glass pipes are a clean-tasting, convenient way to smoke cannabis. Credit: Chris Conrad

Another way of sampling is to roll up several joints, one of each strain to be tested, then take one or two puffs off each one and set it aside, try the next one after a while and so forth, keeping track of which joint was

### Keeping Your Pipe Clean

If a foul-looking oil starts coming out the back of the pipe onto your hands or, worse yet, on your lips or in your mouth, it is definitely time to clean the pipe. That stuff tastes terrible and it's pretty nasty.

To take care of the problem, get out some isopropyl alcohol in a jar, warm slightly (careful, it's flammable) and add rock salt to swish the pipe around in (or in sealed plastic container you can shake up). Remove it from the alcohol, wipe out any residue, dip into some warm soapy water, rinse with clean water and wipe with a clean cloth, then let it sit so any alcohol can evaporate. You can keep the alcohol/salt mixture to reuse, and you don't really have to go through all that except that it helps you better taste the terpenes and flavors of the material. Even though it seems to take forever, remember that time is distended and it's really only been a few minutes.

which (you might want to mark each one so you can tell them apart). It is hard to use hash resin in a cigarette, so it is usually consumed by pipe. The effect of smoking resin in its various forms is stronger but clearer than that of smoking bud. The lighters, papers, ash trays, rolling trays, pipes and other delivery systems offer lots of fun and entertainment for the newbie and are always appreciated as gifts for tokers.

The thing about toking, though, is that you taste the cannabis flavors — but you also taste the smoke.

## VAPORIZERS & VAPE PENS

The smoke taste is why many patients — especially those with immunological issues or who are treating pulmonary or throat conditions — prefer to vaporize cannabis, which renders a vapor up to 95 percent cleaner than

Smoking has been the administration method of choice for eons and is remarkably safe, despite the smoke and unknown contents of street cannabis; while it can trigger short-term bronchitis, no long-term study has ever found a link between cannabis smoke and serious pulmonary diseases like CPD, emphysema or lung cancer. The overwhelming weight of the evidence indicates that joints are much safer than tobacco cigarettes in every way.

Vape pen E-cigarettes don't produce smoke. Credit: VapenCity

cannabis smoke. The Volcano accepts crushed bud and has a special extract mesh as well. By adjusting the temperature, the consumer can determine the potency, heat and cannabinoids or terpenes contained in the vapor. The effect is not unlike the smoked effect, immediate and heady, except that vaporized bud tastes cleaner and sweeter without smoke. Likewise, the bud's effect seems more subtle, and perhaps more similar to the effects of smoking hash.

Vape pens, similar to those used for tobacco and other flavor agents, tend to need extracts that can flow over the heating element to be released. The extracted resin is sometimes sold in a cartridge, or in some cases the entire vape pen is disposable. The effects are quite nice, but the plastic waste is hard to justify. Budtenders should be educated about the benefits and proper use of vaporizers, but should also be educated on the relative harmlessness of smoking for adults with healthy lungs.

## A LITTLE DAB WILL DO YOU

Another rapidly growing market niche, the practice of vaporizing high-purity concentrates on high-temperature plates or "nails" has become quite popular with the heaviest cannabis users. The first time you watch a "dab" may be shocking, but the vapor from a dab rig, like that from a vaporizer, is cleaner than the smoke from a joint or pipe.

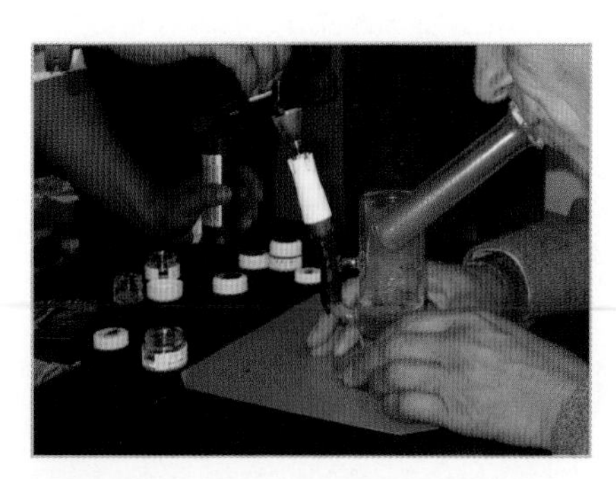

Dabbing on a superheated "nail" instantly vaporizes highly concentrated resin into a single breath; but beware, newbie – it delivers a very strong hit. Credit: Mikki Norris

Dabbing gives the maximum vaporizing dosage; a gob of resin extract is placed onto a red-hot surface and the vapor is inhaled in one breath. Take only a very small hit at first, as most people simply can't handle that magnitude of dose; it can be the equivalent of smoking an entire joint in one breath. Make sure you're sitting down first before inhaling. Seriously. Don't fall over; it can hurt.

## INCREDIBLE EDIBLES

Edibles are wonderful, but they can be tricky. In the hippie days, edibles were just big hunks of chewy brownie with a green cast and gritty texture that we all choked down somehow. An hour later, we were laughing out loud.

My, how things have changed. Many cannabis consumers love the array of delectable treats offered at a dispensary and some simply prefer not to inhale. Edibles come in many sophisticated forms and can be a great way to take your medicine, but in this area the industry has faced early challenges in standardizing the potency of each dose. In early 2013 in Colorado, edibles lab-tested at the behest of the *Denver Post* often were found to contain levels of THC either much higher or much lower than advertised on the label. This is not a new phenomenon, because the digestive process itself slows the onset of the cannabis by one to two hours and metabolizes THC into hydroxyl-11-THC, an eccentric molecule which may either diminish or greatly increase the psychotropic effects. Plus it lasts for six hours or longer, so whatever ride you're taking is going to be a long one.

Fortunately, there are now signs that the industry is cleaning up its act: when the *Denver Post* conducted a follow-up study in 2014, it found much more consistent potency than the previous year. Likewise, the standard single dosage has been lowered to 10 milligrams of THC per unit to prevent over-medication. A lot of people will barely feel that amount, but patients

Make sure you know what dose you're consuming, as one cookie may contain multiple doses. Credit: KCCS

who choose a stronger dosage can take more than one, once they learn their comfort zone. Labeling is key; it's critical for producers of cannabis edibles to clearly identify the fact that it contains cannabis and what its standard THC content is.

Cannabis is also available in beverages, desserts, "green pills," chocolate candies, other candies, and all sorts of other forms too numerous to name here. We urge caution with all of them, but especially the best tasting ones because it's easy to eat too much and then — well, watch out. Less is usually more. And keep them away from kids! It won't hurt them, but child welfare services might take them from you if they test positive for cannabinoids.

## TINCTURES BY THE DROP OR SQUIRT

Often overlooked as a delivery system, cannabis tinctures can be a convenient and easy way to ingest; a small vial containing 30 doses can easily fit discreetly in a purse or pocket, and handy droppers can help patients find consistency. Because tinctures can be highly concentrated, a person has to know the potency before using the dropper, but fortunately any psychotropic effect is almost as fast as smoking, particularly if taken sublingually rather than squirted down the back of the throat and swallowed. Tinctures using terpenes, CBD and cannabinoids other than THC have little psychotropic effect. As with edibles, labeling is a key issue. Make sure to shake tinctures thoroughly before each use, as the resin can otherwise settle and concentrate at the bottom between uses.

Standardized and uniform tincture doses are assured through mass production.

## TOPICALS, LINIMENTS, OINTMENTS & SPRAYS

Sometimes you just want to take care of your physical aches and pains without worrying about getting too "high" or failing an employee drug test; for times like these, topical creams may be the perfect answer. Cannabinoids absorbed through the skin don't cross the blood-brain barrier, so they never reach the endocannabinoid receptors — but that doesn't mean they can't work magic as a localized analgesic, anti-inflammatory and antiseptic remedy. Look for a cream with a base of arnica oil or other deep-penetrating solution.

## PERSONAL DOSING FOR COMFORT & SAFETY

Some people don't feel anything the first few times they toke, but from the "one-hit wonders" to the most redoubtable dabber, every cannabis consumer must find the dose that's right for them. How do you find yours? How do you make sure that your every experience with cannabis is safe and comfortable? For a complete newbie, the answer is simple. Start very slow. Don't be afraid to just say "no more" if you've had enough.

First, inquire about the potency; how strong is the herb or edible? Just remember that it might affect you differently than others, so the answer is a relative gauge, not a certain appraisal of the effect.

Try just one hit and exhale after a few seconds — or try just one small nibble of edible. Wait for it to take effect (see following section for relative timelines). If after about five minutes for inhaled cannabis — about an hour or two for eaten — you don't feel much effect, take a little more. Ease into it; not much cannabis is needed to get a newbie high, in most cases. Just take it slow to be on the safe side, and if it gets too strong, relax and remember: this, too, shall pass. Taking too much can never kill you, but it can scare you enough to call 911, which is both embarrassing and a waste of emergency services. Try not to do that.

Many newbies (and even experienced users) complain that cannabis is too strong in the age of legalization. This has less to do with the national supply of cannabis getting stronger than it does with the legal dispensary system making the potent stuff much easier to find. Whereas in the illicit market, people often want the strongest form of cannabis to get the most "bang for their buck," in the general market, people look for their level of enjoyment. Just as people consume a lot more beer and wine than they do of grain alcohol and vodka, they also tend to ease up on the most potent forms of cannabis. Fortunately, the expanded choices facing cannabis consumers in legalizing states are beginning to recognize that point, by making *less intoxicating* cannabis more readily available, too, at a more modest price. Otherwise try using trim, or even mix in some shake.

For newbies afraid of overdoing it, marijuana that includes cannabidiol (CBD) may be their best friend. While it has non-psychotropic medical benefits, CBD can occupy the same receptor sites as its psychoactive cousin, THC, so cannabis with significant levels of CBD may provide the same medical relief that patients need while mitigating the high that might otherwise interfere with their daily demands as workers, parents, cops or civic leaders. In fact, many consumers of high-CBD/low-THC cannabis report feeling no high at all after taking their medicine, and people who find that high-THC stuff to cause anxiety often find that a combination of CBD and THC is more relaxing.

This is great news for any newbie. If you live in a state where cannabis is legal, browse around Weedmaps.com for retail outlets in your area that have high-CBD cannabis on their menus. If you try and like these less intoxicating varieties of cannabis, it's then relatively easy to go on to try other blends and strains and save the stronger stuff for special occasions.

Most legalizing states are including regulations requiring retailers to advertise laboratory test results for cannabis they sell, including relative concentrations of THC and CBD. If you live in such a state, take advantage of this benefit! Cannabis that tests high in THC (about 15 percent or above) should probably be avoided until you've built up some tolerance; but remember that the *ratio* between THC and CBD is more important than the absolute concentration of either cannabinoid. Thus, cannabis that has 15 percent THC and 15 percent CBD (1:1 ratio) will be less intoxicating than cannabis that tests at 10 percent THC and 1 percent CBD (10:1 ratio), even though it contains more THC. Cannabis that tests 4:1 THC:CBD or less is preferable for newbies.

Chances are that you will not have all that info, especially if you are just sharing a toke with an acquaintance, so just remember, go slow and have patience. Enjoy your cannabis without going overboard.

For edibles, start with about 10mg of THC per dose (the recommended dose by the Colorado government) and if it's too weak an hour later, go up another 5 to 10 mg until you find your level of enjoyment. This opens another point of discussion.

## COMPARE SMOKING WITH EATEN EFFECTS

The timing and duration of a cannabis "high" can greatly depend on how the dose is taken. Consider the following chart:

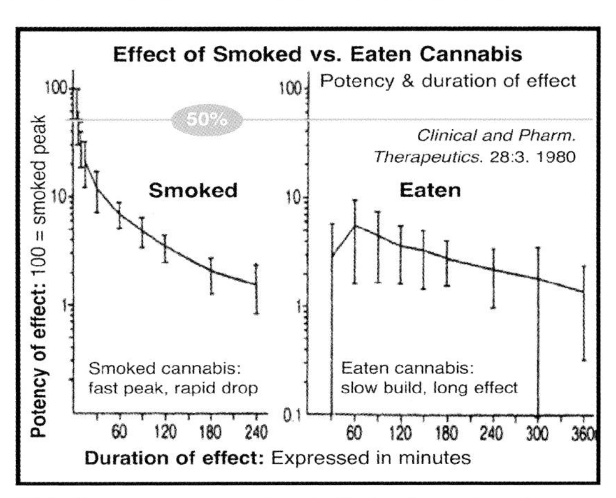


This chart compares onset and effects of cannabis when smoked (l) and when eaten (r). Edible cannabis kicks in later and lasts longer.

What a difference a brownie can make! As you can see on the left-hand side of the chart, smoked cannabis takes effect almost immediately with a short peak and rapid drop; yet on the right, cannabis that is eaten builds slowly toward full effect up to an hour after ingestion before very gradually wearing off over the course of six hours. This is why it's more common to over-medicate from cannabis edibles than from a joint, vaporizer or bong. Newbies who inhale too much usually can tell very quickly and stop (it might just be one puff or dab); in a few minutes the high will diminish to manageable levels. But edibles aren't quite so forgiving because many a naive newbie has found that first bite of cannabis brownie too delicious to refuse a second... or a third... or a fourth. Next thing they know? Four hours of couchlock.

Don't let this happen to you. Until you know your limits with edibles, take just a small bite at a time and wait up to an hour or two between nibbles. It may seem like an incredible act of willpower (and especially so once the famous "munchies" kick in), but hold strong. While eating too much brownie won't ever kill you, it can definitely knock you right out, and has turned some newbies off from ever trying it again.

## APPROPRIATE SET & SETTING

Did you try cannabis at some time in the past and find that it just didn't agree with you? Did you feel wiped out, paranoid, withdrawn, depressed? Did you quit then, figuring that the herb maybe just isn't for you? Don't give up — maybe it wasn't the cannabis at all.

Dr. Norman Zinberg turned the conventional wisdom about drug use on its head with the publication of his monumental *Drug, Set and Setting* in

Cannabis can enhance your experience of nature and the great outdoors. Credit: Kkmd

1980. Whereas most people assume that a drug like cannabis would have a consistent effect for most people and especially for the same person over time, Zinberg showed that the mindset of the consumer and the setting in which the drug is consumed exert an effect equally as powerful as that of the drug pharmacology.

Simply put, set and setting matter. For example, don't use cannabis just because you're in a weird situation and looking for escape; if you begin to consume cannabis with a depressed mindset, it may help or it may only intensify your depression. Similarly, most experienced users wouldn't recommend that a newbie take very much when around unfamiliar crowds, in stressful environments, before taking final exams or arguing a case before the Supreme Court. It will probably be OK, but you may well wish you hadn't eaten that extra brownie.

Don't get us wrong, because sometimes cannabis can lift you right out of a bad mood or a pique of anger or distract you from an irritant; but sometimes these kinds of settings could simply trigger paranoia and lead you to a bad experience. Who needs it? So if you can opt for a better situation, do so.

If you're just starting out with cannabis, it's best to find an environment where you'll be comfortable and at ease. Calibrate your mindset by doing something you'd enjoy anyway, even if you were sober. If you like hiking in nature, wait until you do it on weed. If you like looking at artwork, wait until you do it on weed. If you like listening to or making music, you have to try it when you are high. This herb can work wonders and make you feel like you are experiencing life for the first time, but remember that it can only work with the ingredients you provide.

## SHARING WITH FRIENDS & STRANGERS

People love to get together in a circle, share a joint or a vape and talk of many things. You will probably find this happening to you if you are open to it. Someone will smell the smoke and wander by, and before long, you're sharing with them, too. Or vice versa. It's a great way to meet people. Just remember to practice safe smoking technique. Smoke joints "chillum-style" by forming a fist pipe with your hands, so your lips touch your fingers and not someone else's saliva. If sharing a pipe, wipe the end with an alcohol swab to prevent transmission of germs.

## PUBLIC SAFETY ISSUES

Wherever we go, one question that keeps coming up is: How will legalization affect road safety? We will talk more about this later (see "Cannabis in Society"), but for right now let's just say that when you consume cannabis you are still responsible for the things you do.

The peak effect of smoked cannabis is in the first 15 minutes after toking, so wait at least that long before doing anything that needs too much coordination (you can still play sports and such, of course; we're talking about risky stuff).

If you feel impaired, do not drive. You probably drive too much anyway, so let someone else do it. Experienced tokers usually drive as well as normal; a few drive better when high, because they drive more slowly and cautiously. But newbies may find it difficult to drive smoothly or time their reactions correctly.

Remember that using cannabis by itself is safe, but that it can enhance

> **EVERY NEWBIE NEEDS TO KNOW:**
>
> Cannabis affects different people differently; if you often find yourself zoning out or dozing when you need to be awake and present, change something about your consumption habits. We all have to know our own limits.

### Am I Good To Drive?

1) Before you drive, check your coordination; stand on one foot for a minute and then the other foot without tipping; do finger taps to test your hand and eye coordination. If you can't do that, you shouldn't be driving, whether you are high or not.

2) Tell your "co-pilot" to keep an eye on your driving and let you know if they think you should pull over and, if they say yes, take a break without another hit before you get behind the wheel again.

the negative effects of more dangerous drugs like opiates and alcohol. Know your contraindications; when in doubt, ask a doctor.

Don't use cannabis as either an escape or an excuse. It's a wonderful treat at the end of the work day or week, but a poor excuse for not getting work done in the first place. Many people toke and are productive; you can do likewise.

If friends or loved ones approach you about how you engage in cannabis use, don't assume that it's just more Reefer Madness. If your consumption levels genuinely make others uncomfortable, consider adjusting your lifestyle. It's not hard to cut back on cannabis; all it usually takes is some willpower.

## How To Roll A Joint

Rolling your own joints can be easy once you get the hang of it; and until you do, simple rolling machines are inexpensively available at most head shops and liquor stores.

- *Grind the bud.* This is an all-important first step that is all too often forgotten. You want a fine grind, but not powder. Some tokers use a dedicated coffee grinder, but most hand grinders will do great. You can break it with your fingers but it is slower and less even in texture.
- *Pack it loose.* The ground bud should rest lightly in the paper; if you compress the bud too much the joint will become hard to pull from, especially once resin starts gumming everything up.
- *Tuck the paper.* This is the trickiest step; the idea is to rest the bottom side of the rolling paper (the side without the glue) on the newly formed cylinder of bud, then roll the joint up in your hand or against a surface so that the paper tucks in and rolls over itself. This step may need some practice.
- *Moisten the glue.* Most tokers use their tongue, but a damp sponge works great, too. Dry mouth can be a challenge and tear the paper, so take a sip of water first.
- *Roll it over.* Finish the roll you started by tucking the paper again. Bring the moistened glue down. Find any points where the paper may be sticking up, and lick it down.
- *Use a filter.* These short cylinders, often made from cotton or foldable cardboard, can help you keep the joint's shape as you roll it. Then, when you light up, they create a barrier between your lips and the tar-like resin that can collect at the end. Plus you won't burn your fingers or lips.

Manicured and dried bud

If you are like most adults, you will find it easy to have a positive relationship with cannabis that will benefit you throughout your life.

## Other Ways To Light Up

Not every cannasseur likes to use cigarette lighters to ignite their pipe, joint or bong. Fortunately, you have options:

- *Hemp thread,* when coated in beeswax, provides a sustained low flame — no fossil fuels required. Some seasoned tokers claim that this makes for a cleaner taste.
- *Solar rips,* taken on a sunny day with the help of a magnifying glass, taste amazing. Find the distance from the glass where the concentrated sunbeam comes to its narrowest, hottest point.

# 4 MEDICAL MARIJUANA

There are many ways to consume medical marijuana, including oil extract (top L), kief (center L), pre-rolls (bottom R), tincture (top R) and bud (center).

Cannabis has been among the most widely used of medicinal plants for over 3,500 years. We used it here in the USA from 1850 to 1941 to treat more than 100 distinct diseases or conditions — yet the human endocannabinoid system was discovered only about 20 years ago. Today, scientists hold international conferences to present medical research on the effects of phytocannabinoids and discuss new advances in cannabis therapeutics. Millions of patients in at least 23 states use cannabis with some level of legal protection. The U.S. government holds Patent 6630507 on cannabis medicines. Dispensaries are operating openly in the nation's capital.

Cannabis and all the cannabinoids are non-toxic. All its effects are temporary. It is exceptionally safe; not one single death by cannabis overdose has ever been reliably reported in medical history. Yet, people still go to prison for marijuana — medical or not.

One might hope that the question of whether to use medical marijuana would be a matter for a patient and their physician to decide; however, the hands of big government are inserted deeply in between the two. With the introduction of the federal Compassionate Access, Research Expansion and Respect States (CARERS) Act to allow states to regulate medical marijuana and the widespread public acceptance for medical use, that situation could well change. Medicine is big business, and big business has friends in high

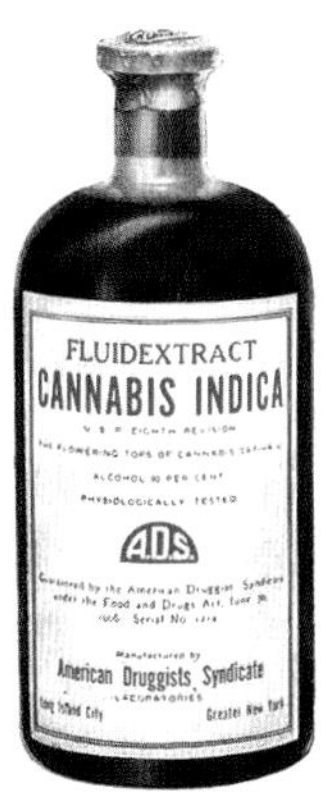


Medical marijuana is nothing new. This cannabis tincture was bottled by a U.S. pharmaceutical company about a century ago.

places; once the federal obstacles come down, market forces could bring a quick revival of cannabis therapeutics.

California allows cannabis use for listed treatments or "any other condition" that a physician approves, but most other states are much more restrictive. People try to stay well within the law when possible, but when you or your loved one's health and quality of life are on the line, you have to set your own priorities.

## THE BASICS OF CANNABIS & WELLNESS

Hempseed is healthy to consume but it has no drug effect. It is a nutritious nut-like fruit that contains proteins and omega acids to bolster the immune system, works as a gentle laxative and may reduce "bad" cholesterol levels. There are many ways to consume hempseed — some as simple as sprinkling hulled seed on top of your favorite dish — and is found in numerous "super food" recipes. Its oil is used in many foods, salves, lotions, and hygiene, health and body care products. So hemp seed is definitely medical; but it's not marijuana.

What, then, is medical marijuana? The basic premise is that the human endocannabinoid system does a pretty good job of of bringing the body into homeostasis and keeping it there; but sometimes it might need help. Cannabis plants come to the rescue with an ensemble of phytocannabinoids, unique terpenes that attach to special receptor sites in the brain and other areas of the body. It seems to work in two ways: stimulating cannabinoid receptors when there are deficiences in the person, and rebalancing or soothing the human system when it is overstimulated. Some of the specific mechanisms remain unknown, and their effects can vary greatly between individuals.

### EVERY NEWBIE NEEDS TO KNOW:

"Cannabis therapeutics is personalized medicine. One size doesn't fit all with respect to cannabis therapeutics, and neither does one compound or one product or one strain." — *Martin A. Lee, director of Project CBD, author of* Smoke Signals: A Social History of Marijuana — Medical, Recreational and Scientific

## SYMPTOM MANAGEMENT & SIDE EFFECTS

Its wide spectrum of direct effects make cannabis beneficial for many diseases, some of which are identified in clinical studies and others best known through "anecdotal" case histories. Cannabis can help control symptoms and slow the development of many ills, bringing safe and effective relief, and significantly improving a patient's quality of life and functionality.

Because the body's responses to certain cannabinoids are well known, they can be used to target specific conditions. For example, most people have heard that cannabis can cause ravenous hunger, called the "munchies." If a person has no appetite, it makes sense that they will benefit from cannabis. Cannabis is both a vasodilator and a bronchodilator, so it can help with asthma and migraines. It can make people sleepy, helping folks with insomnia. It suppresses dreams, which is good for PTSD. It perks up some people, and that's good for chronic fatigue syndrome (CFS). It raises the heartbeat like a stimulant but also reduces stress like a tranquilizer — a fine alternative to pharmaceuticals.

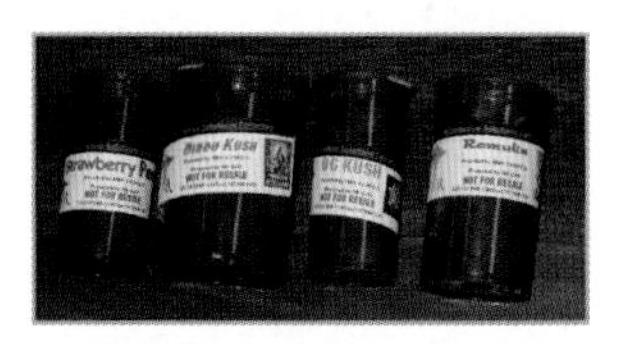

Cannabis is available in many different medical strains. This bud has been packaged in pill bottles for freshness.

Cannabis does all this without toxicity, but it does have side effects: reddening of the eyes; slight temporary increase in the heart rate; cool fingers and toes; dry "cotton" mouth; euphoria; distraction, mild anxiety or, sometimes, dysphoria. Cannabis smoke can irritate the throat and exacerbate bronchitis — just like the smog we inhale every day. Unlike smog, however, cannabis smoke appears relatively safe and does not increase risks of cancer or emphysema. There are many alternatives to smoking, however, so some patients prefer a different means of ingestion. It

| | Effect | | | | | | | | | | | | | | | | | | |
|---|---|---|---|---|---|---|---|---|---|---|---|---|---|---|---|---|---|---|---|
| Compound | Pain | Nausea | Cancer | Spasms | PTS | Neuro | Insomnia | Anxiety | Bacteria | Depression | Diabetes | Inflammation | Psychosis | Fungus | Lupus | ADD/ADHD | Cardio | Psoriasis | HIV/AIDS |
| THC | ☑ | ☑ | ☑ | ☑ | ☑ | ☑ | | ✗ | | | | | ✗ | | | | | | ☑ |
| THCa | ☑ | ☑ | ☑ | ☑ | | ☑ | ☑ | | | | | | | | ☑ | | | | |
| THCv | ☑ | ☑ | | ☑ | | ☑ | | | | ☑ | ☑ | ☑ | | | | ☑ | | | |
| CBD | ☑ | ☑ | ☑ | ☑ | | ☑ | | ☑ | ☑ | ☑ | ☑ | ☑ | ☑ | | | | | ☑ | |
| CBDa | | ☑ | ☑ | | | | | | ☑ | | | ☑ | | | | | | | |
| CBN | ☑ | ☑ | ☑ | ☑ | | | ☑ | | ☑ | | | ☑ | | | | | | | |
| CBG | ☑ | | ☑ | ☑ | | ☑ | ☑ | | ☑ | | | | | | | | | | |
| CBGa | ☑ | | ☑ | | | | | | ☑ | | | ☑ | | | | | | | |
| CBC | ☑ | | ☑ | | | ☑ | | ☑ | | ☑ | | ☑ | | ☑ | | | | | |
| Anandamide | ☑ | | ☑ | | | ☑ | | ☑ | | ☑ | | ☑ | | | | | | | |
| Borneol | ☑ | | ☑ | | | | | | ☑ | | | ☑ | | ☑ | | | ☑ | | |
| Caryophyllene | ☑ | | ☑ | | | ☑ | | ☑ | ☑ | ☑ | ☑ | ☑ | | | | | | | |
| Eucalyptol | ☑ | | ☑ | | | ☑ | | | ☑ | | | ☑ | | ☑ | | | | | |
| Humulene | ☑ | | ☑ | | | | | | ☑ | | | ☑ | | | | | | | |
| Limonene | | ☑ | ☑ | | | | | ☑ | | ☑ | | ☑ | | ☑ | | | | | |
| Linalool | ☑ | | | ☑ | | | ☑ | ☑ | | ☑ | | ☑ | ☑ | | | | | | |
| Myrcene | ☑ | | ☑ | ☑ | | | ☑ | | ☑ | | ☑ | ☑ | ☑ | | | | | | |
| Pinene | ☑ | | ☑ | | | | | | ☑ | | | ☑ | | | | | | | |
| Terpinolene | | | ☑ | | | | ☑ | | ☑ | | | | | ☑ | | | ☑ | | |
| | | | | | ☑ | Promising research | | | | | | | | | | | | | |
| | | | | | ✗ | Contraindicated | | | | | | | | | | | | | |

The many and varied cannabinoids and terpenes in cannabis have a wide variety of effects. This chart summarizes the known or researched benefits of just nine cannabinoids, the endocannabinoid anandamide and 10 common terpenes. An 'X' means contraindicated.

is this lack of serious side effects that makes cannabis so desirable among patients.

> **EVERY NEWBIE NEEDS TO KNOW:**
>
> "Topicals are very economical and effective for pain from arthritis, injuries, migraines, etc. Since they do not enter the bloodstream, they do not cause unwanted psychological effects."
> — *Danielle Schumacher*

## CANNABIS WORKS GREAT, BUT NOT ALWAYS

For some people, cannabis is like a miracle cure. For most people, it is as good or better than other drugs, without their side effects, while for others it may offer no benefit. Of course, no drug works equally well for all people in all circumstances. Neither do all strains work equally well in treating specific problems. For example, a variety that reduces nausea and stimulates appetite may not be as effective at controlling aches, pains or insomnia. Not everyone will tolerate its side effects, either.

Only certain strains of cannabis plants produce enough tetrahydrocannabinol (THC) or cannabinol (CBD) to be used for medical marijuana. Effectiveness is linked to dosage. Some patients find that small amounts suffice, while others need heavy, ongoing doses to function or suppress symptoms; patients who receive their medical marijuana from the federal government average 10 cigarettes per day. Fortunately, it's not a competition.

This Harlequin bud is high in CBD.

The advantage of smoked or vaporized cannabis is that the effect is felt almost immediately, so it is a simple matter to stop as soon as the desired subjective effect is achieved. The subjective effect of cannabis helps people titrate their dosages; if the effect of being high is overwhelming, they step it back a few notches next time or use the small dose instead of the larger dose.

A canister containing medical marijuana

## CHOOSING A DOSAGE & HEALTH REGIMEN

There are a wide variety of conditions for which cannabis can provide relief; your treatment regimen will be determined by the intensity and duration of your symptoms. If the problems occur at night, for example, edibles would be better because they are effective for longer time periods. If there is a crisis like a patient going into seizure, inhaled cannabis is almost immediate in bringing relief and is much easier to administer. If there is an ongoing issue, as with patients trying to reduce continuous seizures, for example, a small oral dose of FECO oil extract ("phoenix tears" or "RSO") is often used, while those who are fighting off late-stage cancer tumors find that much larger oral doses are required.

First look at what specific symptoms need to be treated that respond to cannabis, then see if it has any negative effect that contradicates its use. That will help you to identify the appropriate form, dosage and means of ingestion. Start with a low dose and work your way up. Personal research with the advice and approval of a physician is the safest way for any given patient to determine its potential for them. (Warning: most doctors are not trained about medical marijuana and many tend to be biased against it because of their own relationships with pharmaceutical drugs, so you have to educate your doctor or find the right doctor to work with).

### Cancer Cure Claims Court Controversy

You might have heard that cannabis oil can cure cancer, but this claim is controversial. On the one hand, there are no clinical trials showing that it works and no money to do such trials; on the other hand, there have been measured case studies of tumors dying and studies showing that THC kills cancer cells under certain conditions. And then there are documented case studies that show it working in dire situations where nothing else had worked.

The big concern of doctors we spoke with is that some patients choose to forego treatments that are known to work and rely solely on cannabis. They don't want you to die. No medicine works for everybody, not even cannabis. Talk to your doctor; more physicians are now willing to talk about medicine in terms being a complementary regimen of cannabis, healthy lifestyle and conventional medicines for maximum benefit. That is, to use cannabis in conjunction with rather than in place of current medical treatments.

Here are some common uses for medical marijuana:

**Analgesic, pain relief:** Chronic pain is the most common ailment cited by cannabis patients. It helps diminish pain and makes it more tolerable, so patients can partly or totally wean off pharmaceutical pain killers. Both marijuana and some hemp strains seem to be effective, since cannabidiol (CBD), like THC, seems to have a major analgesic (pain lowering) effect. Neuropathy and neuralgia respond well; acute injury pain typically gets less immediate relief, but sufferers often find that cannabis can diminish the intensity and distract them from their pain. But perhaps the herb's greatest

THC, CBD, other cannabinoids and terpenes work together in an entourage ensemble effect.

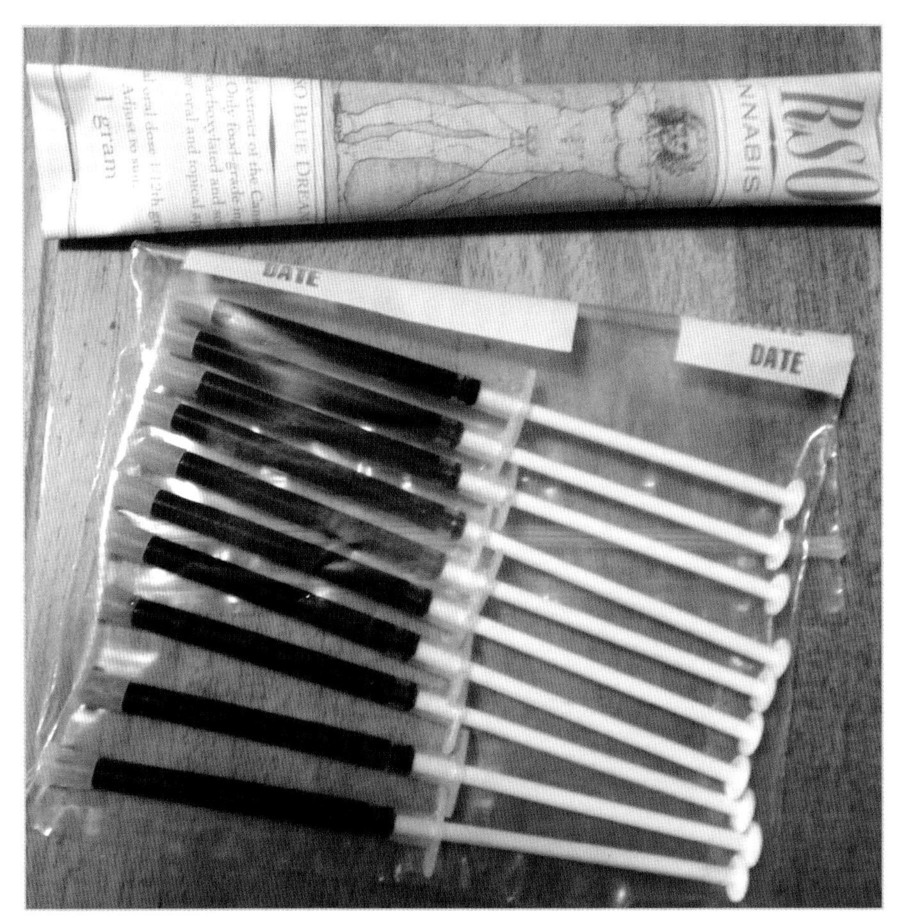


Full Extract Cannabis Oil (FECO), AKA phoenix tears or RSO, is a thick oil that is ingested orally. It is recommended only for patients who need the most potent treatment. Credit: Barry Brilliant

benefits are synergistic; working powerfully with opiates and other drugs, cannabis empowers pain patients to reduce their dosages of prescription drugs that have adverse side effects. Not all pain responds to cannabinoids, but some of the most long-term and troublesome cases do.

**Anxiety, heart disease:** Anxiety-induced stress is a major contributor to heart disease, the No. 1 killer in America. Cannabis speeds up your heartbeat in a way similar to mild exercise, as it promotes relaxation, reduces mental agitation, anger and anxiety, and lends a sense of humor. It can lower blood pressure, provide a welcome distraction from worries and help deepen the appreciation of life. Contraindication: When fast heartbeat poses risk.

**Arthritis, inflammation:** Attention senior citizens: Eating or smoking cannabis helps control joint pain, reduce inflammation and improve mobility. A traditional treatment for rheumatism and arthritis is to soak cannabis leaves in rubbing alcohol and wrap them around the sore joints to reduce swelling and pain. Today, there are many sources and varieties of cannabis-infused creams, lotions, liniments, salves, etc., on the state-legal markets.

**Cancer, AIDS/HIV:** Cannabis reduces the gut-wrenching nausea caused by chemotherapy (and radiation therapy), while stimulating the appetite to help patients eat and combat excessive weight loss and cachexia. It reduces pain and helps cancer patients sleep and rest. It often raises the patients' spirits and mood, improving their will to live and hope for recovery. Direct application of THC in vitro shows promise as a tumor-killing or reducing agent and also a killer of the herpes virus. Some patients who are fighting

Just being around growing cannabis plants is enough to make some cancer patients feel better.

off late-stage cancer tumors report that larger oral doses up to 1 to 3 grams per day of high-THC phoenix tears or RSO oil may be required to reverse growth. You have to work up to that — take small amounts, an eighth or 10th of a gram, and increase it over a period of time. Skin and even brain cancers have been reversed by direct application of oil to the tumors.

**Epilepsy, movement disorders, MS, seizures:** Cannabis is neuroprotectant and can calm down overactive nerves, thereby alleviating seizures that may be caused by a deficiency of natural endocannabinoids. Multiple Sclerosis is characterized by increasing neuropathic pain and degenerative loss of muscle control in two forms: involuntary movements (spasms) and an inability to move (ataxia). Cannabis helps improve movement affected by each of these, while reducing or stopping the pain and related depression. Cannabis provides similar relief from involuntary spasms by sufferers of Parkinson's disease. The CBD-only oils made and used in a manner similar to phoenix tears have dramatically helped in the pediatric treatment of children suffering from crippling seizures, as reported by Dr. Sanjay Gupta on CNN. A combination of CBD with THC and other terpenes might be preferred.

**Migraines:** Cannabis is frequently used to treat migraine headaches. It helps reduce light sensitivity, nausea, vomiting and pain, and can be consumed regularly to prevent attacks from occurring, or to use as needed to reduce the severity of an acute headache. Stress-induced headaches can also be mitigated.

**Glaucoma:** Glaucoma is a common problem and a leading cause of tunnel vision and blindness. Surgery poses severe risk to the eyes, and pharmaceuticals promise dangerous side effects, such as liver damage. Most sufferers could benefit from cannabis, which reduces intraocular pressure (IOP) in the eye caused by a buildup of ocular fluid. The exact mechanism is unknown, but regular cannabis use can often halt this painful progressive vision loss by lowering the fluid pressure within the eye. When acute symptoms appear, toking cannabis can stop an attack.

**EVERY NEWBIE NEEDS TO KNOW:**

"When you use cannabis, always know what you are taking and, when possible, only use tested product that tells you the dose of both THC and CBD in milligrams. Never ingest orally without knowing the dose of THC. Read the label!" — *Dr. David Bearman, M.D.*

**Mental health, Alzheimer's, depression, PTSD:** Cannabis subjectively makes you feel good and forget about your cares for a while, enhances sensory experiences such as enjoyment of music and art, and has long been regarded as a mild aphrodisiac. It releases new ideas, helps with the appreciation of the little things in life and gives people hope. It can stimulate inspiration and critical thinking, increase motivation and reduce malaise such as chronic fatigue syndrome (CFS). It is an anti-depressant and helps people with attention deficit (ADD/ADHD) to better focus and concentrate. It can stabilize bipolar mood swings and may also help with memory, such as with Alzheimer's and senility. Studies on veterans show it helps reduce sleeplessness, nightmares and rage caused by PTSD. It is a viable substitute for mood enhancers such as Demoral, Valium and morphine, but much safer and less habit forming. Contraindications: Possibly in schizophrenia, whenever cannabis triggers subjective paranoia or dysphoria instead of euphoria.

**"You look too healthy for medical marijuana syndrome":** When we look healthy, people assume that we are healthy, and when an apparently able-bodied young person has a doctor's note, people may assume that they "just want to get high." However, a person does not have to look sick to be sick. Pain is invisible. Mental illness is not visible to the naked eye, and a patient appears more healthy and happy if cannabis brings them relief. And besides, if taking medicine doesn't help patients feel better, what is the point of taking it?

**"... And any other condition for which cannabis brings relief":** It is up to the patient to make their own determination with a physician as to whether cannabis is the right medicine for them. Politicians and police are not doctors, nor are prosecutors and judges. Many doctors know that it is hard to find effective and truly safe remedies, and that many of their patients often fight with nurses and caretakers because they refuse to take their medicine. When it comes to cannabis, however, people are usually more than happy to take it, and it puts them in a more congenial mood. If you are a caregiver or manage a seniors' facility, you should be trying to get your patients approved to make everyone's lives easier.

**EVERY NEWBIE NEEDS TO KNOW:**

"When you find out how very useful cannabis is, you may find it hard to believe we ignored and suppressed it for so long because of earlier generations' fears of losing their minds. But it's true."
— *Sunil Aggarwal, M.D., PhD*

Juicing raw cannabis. Just under 100g fresh leaves and flowers (top) juiced (middle) and strained comes to just over four fluid ounces of juice. It tastes a bit like wheatgrass juice. Drink it by the shot or mix with apple juice. Some people just add a handful of cannabis leaf and flower to their smoothie to get both cannabinoids and fiber.

## SPECIAL CONSIDERATIONS

Juicing is a process using the fresh plant to get phytocannabinoids in their natural acid form. It is not psychoactive and is very limited in its availability. Its therapeutic value is still being researched, but it definitely has no illicit market value. Some growers in California produce it for their patient base but it must be used immediately or kept frozen until use, making it impractical for most people except for growers to maintan a supply. The upper leaves and buds are run through a juicer and the fiber is strained out. About a hundred grams converts to about a half-cup of juice. For a milder experience, try scattering a handful of leaves and buds in a blender with water and some fruit. That way you also get some dietary fiber in the blended drink.

CBD strains of cannabis look good, have medical value and a psychoactive interaction with brain chemistry, but they do not have the subjective psychotropic effect of making people high, so they have little value as street marijuana.

Molecular compounds can be isolated and extracted from the plant, as THC and CBD are by GW Pharmaceuticals' product Sativex (not available in the USA yet) or synthesized like THC for Marinol.

## This Is Your Brain On Head Injuries

One of the most urgent applications of medical marijuana is in the field of sports medicine, especially as professional leagues seek out solutions to the plague of chronic traumatic encephalopathy (CTE), which has ravaged the National Football League. This tragic disease causes an inexorable degeneration of the brain in athletes and others who endure multiple head injuries over time; it's heartbreaking for fans to watch their favorite players gradually waste away.

Fortunately, there is promising evidence that the timely use of cannabis could prevent this terrible disease. Raphael Mechoulam, one of the world's foremost experts on medical marijuana, has noted "the endocannabinoids... as well as some plant and synthetic cannabinoids, have neuroprotective effects following brain injury." (Mechoulam et al, "Cannabinoids and brain injury: therapeutic implications," *Trends in Molecular Medicine* 8:2 February 2002).

**CHART: Summary overview of marijuana's effects**

The endocannabinoid system, which modulates other body systems, includes the CB1 and CB2 receptors that, when activated, lead to a chain of physical and subjective effects:

- Lifted mood & enhanced sense of well-being, activated by a stimulant effect followed by relaxation & reduced stress.
- Analgesic. Migraines & seizures are prevented. Alleviates multiple sclerosis (MS) & chronic, neuropathic pain.
- Synergistic effect with opiates and other drugs. Potency and effect vary. May trigger drowsiness, distraction, paranoia or anxiety.
- Eyes redden and dehydrate. Intra-ocular pressure (IOP) lowers.
- Ringing in ears (tinnitus) becomes less prominent.
- Mouth dehydrates, appetite is stimulated, flavors are enhanced. Nausea and vomiting are reduced.
- Inhaled cannabis can trigger cough. Bronchodilation improves oxygen intake. Smoke can irritate mouth, throat and respiratory system; vaporization, oral ingestion and other precautions mitigate this effect.
- Pulse and heartbeat accelerate.

- Bronchia dilate, as do alveoli and blood vessels. When inhaled, cannabinoids travel through the lungs and cardiovascular system into the bloodstream and directly to the brain.
- Mild aphrodisiac enhancement, relaxation.
- Joints are soothed, arthritis eased. Works when taken orally or applied topically.
- Muscles relax. Vasodilation carries blood more quickly, lowers body temperature.
- Fatty tissues collect inert cannabinoids, dispose via urine, hair or feces.

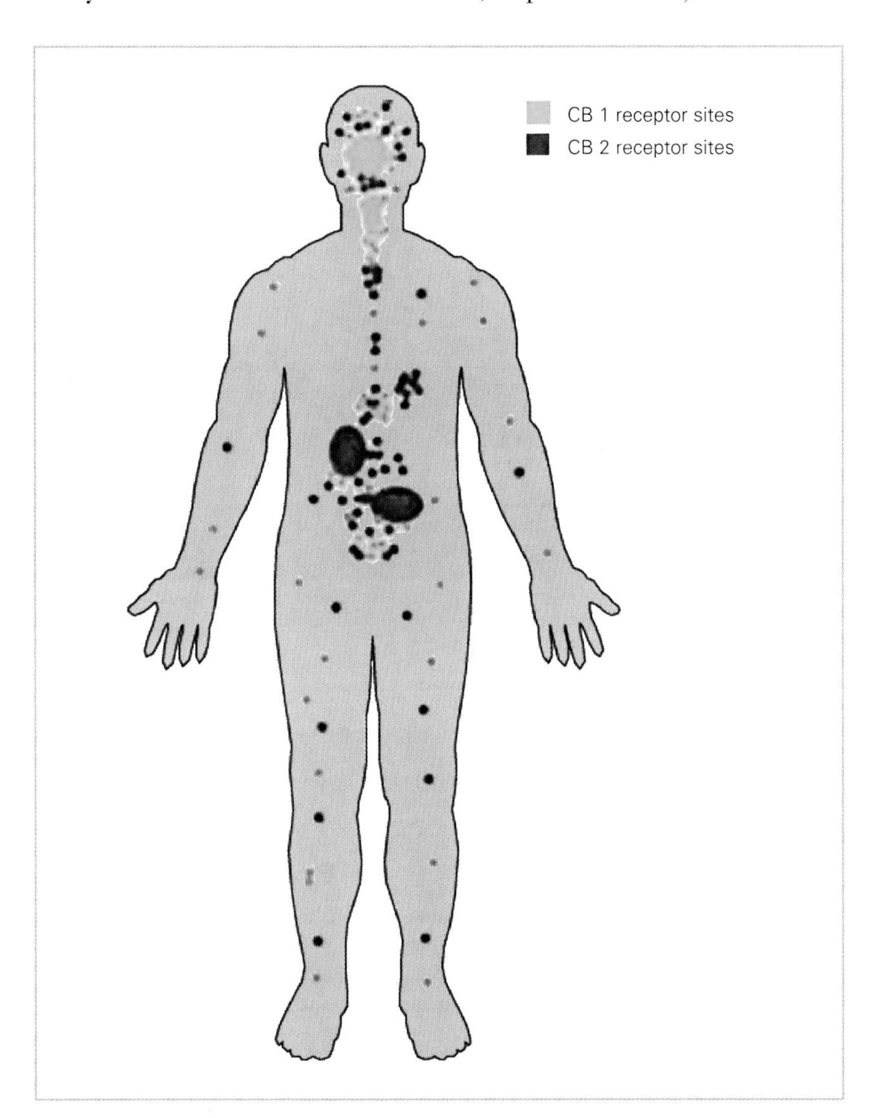

# 5 THE EXPERIENCE OF GETTING HIGH

There is no simple description of the cannabis experience because, truthfully, it all depends — and on more factors than you might think.

Whereas the most popular psychotropic drugs in the U.S. have a single active ingredient with a relatively predictable effect (think alcohol or caffeine), cannabis has several. THC provides most of the psychotropic effects, but it is in the context of numerous related compounds (i.e., cannabinoids and terpenes) that can significantly alter a consumer's subjective experience, depending on their combination and relative concentrations, as well as the user's personal tolerance level. Cannabis highs are the result of the substance consumed, but also strongly influenced by the consumer's state of mind (set), the physical environment (setting) and the kind of activity the user may be engaged in.

Be forewarned: Some people don't care for cannabis' subjective effects at all, but they represent a relatively small slice of the population. And about 20 percent of cannabis consumers like it so much that they use it almost every day, sometimes throughout the day. Some are "wake and bakers" who like to start the day getting high, while others wait until 4:20 or late into the evening to take a toke. Most tokers don't do it every day. Some do it rarely. The majority of people who try it like it well enough to use it occasionally or on weekends and special occasions, but not daily. Which group you'll ultimately belong to is difficult to predict until you've tried the herb to see how it best fits your personal lifestyle.

## THE ENHANCER

Tetrahydrocannabinol, THC, is the main psychotropic compound in cannabis. It exists in acid form in the plant and must be decarboxylated, usually by heat such as smoking or cooking, to have its psychotropic effect.

Much of cannabis' effects depend on what else you may be doing at the time, because whatever feeling an activity or environment inspire while you're straight will be greatly amplified, or enhanced, by cannabis. It has been described as experiencing life for the first time ... again. As Jon Stewart so memorably asked in the film *Half Baked* – "have you ever looked at the back of a twenty-dollar bill... *on weed?*"

This effect of amplified experience is so common that it often will stultify stoner stereotypes. Focus, motivation and concentration can all be greatly enhanced, to the point that physicist Carl Sagan sometimes puffed on a joint before tackling a particularly thorny math problem, and musicians like Bob Marley and The Beatles enhanced their creativity with cannabis before writing some of the world's most profound lyrics.

Some stereotypes do have a basis in truth, though. Food really does taste way better; making love is more relaxed and sensual; and no, you really haven't ever heard *Dark Side of the Moon* until you've heard it while high. So much of cannabis' effects are simply enhancements of your everyday experiences (more on this below).

## IS CANNABIS AN ESCAPE?

Because of the bias of observers and the defensiveness of the consumer, this can be a thorny issue. Many users of alcohol or other drugs find that their use temporarily helps them to "escape" their situation, but for reasons outlined above, that often doesn't work with cannabis — it usually works as an enhancer or through distraction rather than as an escape.

Cannabis is not generally considered addictive in the sense that alcohol, sugar or cigarettes are; but heavy users can become psychologically dependent on it if they try to use it to escape their woeful situations instead of dealing with their circumstances more directly. Smoking may make them feel better for a short while, but in the long term it can enhance some of the negative feelings they felt in the first place, prompting them to smoke even more cannabis. At some point they may have to realize that they are not using cannabis in a healthy way; time to either back off or stop. Some people need help and should get it.

To be fair, though, cannabis can also help you depersonalize from your situation so that you can look at life more objectively. If you felt angry, it may calm you down; if you felt someone did you wrong, it might help you look at the situation from a different point of view and not take it as personally; if you are so deep in the middle of a situation that you can't judge it clearly, cannabis can give you the distance you need to face reality and the calmness to accept the facts. With any luck, it also gives you the creativity to find better answers to your problems.

## INSIGHTS & CREATIVITY

While the pharmacokinetics of *how* it happens are still not well understood, most cannabis users agree that use of the herb significantly increases their creativity and innovation while under its influence. This effect is so pronounced, in fact, that some of the most celebrated authors and poets of 19th century Paris gathered together regularly for a salon called the *Club du Haschichins*, in which well-known authors like Baudelaire and Dumas consumed edible cannabis treats spread on toast before discussing high-minded literary ideas and techniques. The Beatles credited the emergence of *Sgt. Pepper's Lonely Hearts Club Band* as being inspired and influenced by cannabis.

Just beware that not all insights gleaned while under the influence of cannabis hold up under sober scrutiny, just as the things you couldn't stop laughing about last night may not even seem funny today. However, sometimes your "brilliant note" might actually be a stroke of genius, like at least some of the notes The Beatles wrote while they were high with their friends.

Also, try to remember those insights until you can reconsider them.

## THANKS FOR THE MEMORIES & FOR THE LAPSES

Cannabis' relationship to memory is complex. For most users, long-term memory is unaffected; you won't forget your anniversary or your social security number, even if you use cannabis every day. But for sufferers of post-traumatic stress, there is tantalizing evidence that regular use can help consumers forget their most traumatic memories while leaving happy memories unaffected. This research is still in its early stages, but could explain why PTSD patients seek out cannabis at such a high rate.

Short-term memory is a different story; nearly everyone will experience temporary disruptions to this side of memory if they toke a large enough joint. Don't be surprised if you find yourself interrupting your own funny story, suddenly distracted by a profound tangential thought, and asking, "now, what was I just talking about?" Not to worry; this effect lasts only as long as the high itself, and your memory functions will return to normal. If someone reminds you what you just said, your memory of the whole event is still there. If you apply some discipline and retrace your mental tracks, you can easily take control of this — but there goes some of the fun of getting high and being silly.

By the way, people who have never smoked cannabis do that all the time, as well, and nobody gives them a hard time about it; but if you're high, people judge you for it and blame it on pot. It's really a double standard, so tokers have to be better at remembering than a straight person is, just to be regarded as the same. In 1998 after smoking numerous joints, we were asked to memorize four random words to test our memory retention. For the record, those words were "heretofore, perfect, illuminate, and Constantinople." It's been 17 years now and we still remember.

## JUDGMENT & PUBLIC SAFETY

Alcohol is infamous for spurring lapses in judgment in its users; almost anyone who has consumed any significant amount of that drug is familiar with its propensity to lead to bad decisions. In response, some people feel hesitance about creating an entire commercial industry around cannabis, and stoke fears that legalization unleashes yet "another drug" that will cause lapses of judgment among its users.

Fortunately, there is a big difference between cannabis and alcohol in regard to public safety. Whereas an adult under the influence of alcohol may tend to overestimate their own ability to drive a car and jump behind the wheel, for example, a cannabis user who feels too "stoned" is more likely to ask someone else to drive. Drunks drive fast; tokers drive slowly. Drunks run through stop signs; tokers sit at a stop sign waiting for it to turn green. Drunks drive the wrong way down the freeway ramp and cause a wreck; tokers miss their off ramp and get lost trying to find their way back.

The biggest danger of mixing cannabis and driving is when you mix in alcohol, too. If you are drunk, you might feel like a toke will sober you up, but that's the bad judgment of alcohol at work. Your stoner sense should tell you otherwise. This effect applies generally: Do not drive drunk, no matter what other drugs (cannabis or even coffee) you are using to counteract it. For most tasks that could be affected by cannabis use, the user is able to gauge their impairment and adjust behavior. This can be a good instinct to listen to; if you feel you're too high to drive, you probably are.

When it comes to important decisions, cannabis is more likely to make one hesitate and reconsider than to act hastily. It can make you pause, look at things from various perspectives and consider possible consequences before you act, or just to wait until you are no longer high to make any final decisions.

## TIME TO THINK ABOUT THINGS

One of the most common effects of cannabis use is a distortion of the user's subjective experience of time. Most cannabis users experience this as time dilation — an experience that seemed to last hours really took only a few minutes. Plenty of cannabis users find themselves getting so lost in a beautiful song that goes on and on that they find themselves incredulous to discover that the track is still only three minutes long. Cannabis can make an experience seem to last longer and therefore be more fulfilling. Some people feel that if they go most of the day without smoking and then smoke in the evening, it's almost like getting an extra day.

While this effect can greatly enhance the user's enjoyment of music, sex and nature, the time dilation of cannabis highs can also lead to slower reaction times, and make something that is tedious and boring seem even more so. It can make it hard to follow complex threads of information and cause you to get distracted, and then suddenly realize you were not paying attention. So for some it is not a good choice in those situations. On the other hand, if it taps your curiosity and helps your concentration, this extension of time adds to the fascination of these complexities. In the end, it's all subjective.

## SEX & SENSUALITY

Ever since buttoned-up Victorian times, Western observers have noticed a connection between the use of cannabis and the enjoyment of sex; Hindu authors of the Kama Sutra had it figured out thousands of years earlier. Some consider the herb a mild aphrodisiac as the heartbeat picks up, triggering excitement; blood flow is increased to the organs; and while under its influence, a wide range of sensual experiences become more enjoyable. Some women (and men) who have trouble experiencing orgasms have reported that use of cannabis helps them to take their time and ease into a deeper relaxation where they can access that intense pleasure, which seems extended for a longer period of time — and just enough mental distraction to keep things interesting.

Sensual enhancements can be cumulative; soft candlelight and romantic music become quite enjoyable all on their own while under the influence of a gentle cannabis high, but when combined all together the effect can be mind-blowing. It can make sensuality and foreplay just as fun as the sex itself; but just remember that, while cannabis can help to ease inhibitions in the bedroom, this is no excuse for not practicing safe sex.

## FOOD, FLAVORS & FEELING FAMISHED

Almost everyone who tries cannabis agrees that use of the herb can trigger a ravenous appetite and greatly enhances the enjoyment of each morsel of food; however, the connection between cannabis and food goes much deeper.

Patients with nausea or reduced appetite from cancer treatments, HIV/AIDS wasting syndrome or gastrointestinal disorders overwhelmingly report a positive effect from using cannabis, which often stimulates their appetite more effectively than any other treatment. Others with eating disorders such as anorexia or bulimia report that the use of cannabis can help them break the destructive cycle of these conditions and restore a healthy relationship with food.

But while cannabinoids — and THC in particular — are known to stimulate appetite, other cannabinoids like tetrahydrocannabivarin (THCv) have been identified as powerful appetite suppressants, leading some researchers to speculate that this lesser-known compound could be one of the most promising diet drugs ever identified (see "Cannabinoid Chart," Chapter Four).

Cannabis can make each flavor stand out in a dish in an amazing panorama of olfactory delights; it may send you binging on chocolate candy, eating so fast you lose track; it may send your appetite packing and leave you uninterested in food. One way or another, don't be surprised when you find your attitude toward food being adjusted by cannabis — which may not be a gateway to heroin, but is definitely a doorway to Doritos™. This is what makes munchies such an important ancillary business to the cannabis industry.

## THINKING & CONSCIOUSNESS

In his 1972 treatise *The Natural Mind*, Dr. Andrew Weil wrote about the differences between "straight thinking" and "stoned thinking" — while straight mindsets get caught up on materialistic and intellectual trains of thought, stoned thinking helps us break out of such limited heuristics.

Liberation has its benefits. Consumers may feel hope, happiness or euphoria — a sense of well-being and contentment. Problems seem less severe and pressing. Mundane things suddenly seem more interesting, alive and rich in details you never noticed before. Simple words achieve a new level of profundity. Stoned thinking becomes free streaming; it may change direction in a moment and has been compared to the indirect move of the knight on a chessboard, as opposed to the direct linear moves of the rook or bishop. It's like switching among various frames of mind, seeing things from different points of view. It may take you off on tangents as you follow a peripheral train of thought. Other times people just get the giggles and laugh about anything, no matter how silly or ridiculous — the more absurd, the funnier. For some, however, thinking may be "foggy;" less clear or focused, more observational than critical. Some people become more passive or self-conscious and talk less, especially when on stronger doses. Others become more spontaneous, sociable and whacky.

You may realize that something that has really been bothering you is actually not such a big deal after all, or that you had overlooked something that is key to a problem you've been trying to solve. You may identify or imagine relationships between things that you had not grasped before. A profound metaphor may reveal itself with rich implications for your life and destiny, enhancing your sense of spirituality. This aspect has led to cannabis being used as a sacrament in many of the world's religions throughout history. There are Rastafarians and a variety of Christian, Hindu, Islamic, THC Ministry, Church of Cannabis, animists and other religious denominations that revere cannabis as a sacred plant to be used as a sacrament; these are categorized as cantheism religions. Bible scripture is cited to support its use, particularly by branches of the Coptics and some Essenes.

## PARANOIA & ANXIETY

A common adverse effect of using cannabis, especially for novice users trying a variety high in THC and low in other cannabinoids, is the triggering of feelings of paranoia and/or anxiety. Much of this effect can be attributed to simply living under prohibition. As California NORML's Dale Gieringer put it, when a cannabis user is arrested every 45 seconds in the USA, is it any wonder we're paranoid? Yet this effect cannot be attributed to laws alone; even in legal states, novice users report feeling anxiety and other unpleasant sensations under the influence of THC. This is often related to the speeded heartbeat, so being aware of this side effect can help to control possible panic; no, you are not agitated, your heart is beating fast in response to the cannabis.

If this happens to you, one of the best preventatives is to seek out cannabis with lower THC potency and/or higher levels of CBD, which is a potential anti-anxiety medication. Not every outlet offers mild low-THC or high-CBD cannabis, but as laws continue to change and market demands shift, this pleasant wonder drug is becoming more and more available.

Another technique to mitigate anxiety is to avoid high doses of edibles, which can more easily lead to adverse effects than smoking or vaping because the hour or longer time lag before its effects take hold. If this happens to you, just relax. Try drinking some lemon water, getting fresh air or, if even that feels like too much, simply lie down, relax and let your mind float downstream. This, too, shall pass.

## RELAXATION & TIREDNESS

Many people report that using cannabis helps them to relax and brings down their energy when they are stressed out. Others report that it actually lifts them up and gives them bursts of energy. What is going on here?

As noted before, not all cannabis is the same. Depending on the strain and the techniques used to cultivate it or how old it is and how it was stored, various products may have very different cannabinoid and terpene profiles, which will lead to different effects.

Likewise, not every cannabis user is the same, and even the same user can change over time in a matter of hours or of years. Dr. Raphael Mechoulam postulates that stimulation of the body's endocannabinoid system may tend to return the body to homeostasis — the normal baseline of human functionality to which the body returns during rest periods. That could explain why people who are wired on coffee often find that the herb "brings them down" while those who feel tired before using it feel a lift of energy. Cannabis responds to whatever state your body may be in at the time. Or you might be one of those sleep-deprived people who really needs more rest and the cannabis is just making you face the fact that you need a nap.

## OVERDOSE? NO, JUST OVERMEDICATED

First the good news: In over 12,000 years of human interaction with the plant, there is not a single recorded case of anyone ever dying from cannabis overuse. The consensus among modern scientists is that a lethal cannabis overdose is physically impossible, because the endocannabinoid system simply offers no opportunity for cannabis to interrupt vital functions like breathing and heartbeat.

### Why Didn't I Get High?

Many newbies report that the first time they tried cannabis, they didn't feel any different. This is pretty normal. The experience of cannabis, like those of most drugs, depends on the set and setting as much as on the drug itself. The brain has to learn how to get high by finding the right combination of mindset and environment; after that, the experience happens more automatically.

If you've tried cannabis but haven't felt high yet, try focusing on your breath and the way your body feels as you inhale or ingest. With each exhalation, relax a little more into your body. Soon you'll notice a wonderful feeling come over you.

Now, the bad news: It only feels that way. Novice users who take way too much may experience an intense cannabis high that can be disorienting; the heart pounding, the mouth pasty, the body heavy and tired, head spinning, waves of dizziness or nausea. It's nature's way of saying, "enough with the cannabis for now." Thus, the most common symptoms of cannabis overdose include paranoia and even panic, or alternatively, sudden feelings of sleepiness (another reason why novice users shouldn't try to drive until they feel comfortable and the effects begin wearing off).

Fortunately, even the most extreme overdose wears off in time with no lasting effects except a distaste to repeat it. Generally, there's no such thing as a cannabis "hangover," although users who take too much the night before may start slowly and feel sluggish the next morning. Cannabis will let you know when you've had enough. The best thing to do is to learn your limits and stay within them, so you can handle whatever the day requires of you. You want to have a pleasant, positive experience with the herb and to get the most out of your experience.

## SOCIALIZING & MAKING FRIENDS

The cannabis user's enjoyment of social space can be highly variable. Many users report feeling more relaxed and less inhibited around friends, but others feel more introspective and withdrawn — especially around strangers and big crowds. Cannabis certainly isn't the social lubricant that alcohol is, but it's a lot of fun in its own way.

Moreover, cannabis is generally shared among a group of people, and wearing a cannabis leaf logo design will often get you an invitation to meet new people and share in the cannabis experience, sitting or standing in a small circle, talking with people you didn't even know before. It creates an immediate bond and gives you plenty to talk about: the cannabis, previous experiences with it, story telling, jokes, food, shared love of art, philosophy, music, etc. Cannabis opens the door to make new and interesting friends everywhere you go. Just watch out for narcs (and don't be paranoid). Also, bring plenty of lighters, as they tend to disappear from time to time.

## READING WHILE HIGH

When it comes to reading, once again, your mileage may vary. Many experienced users report that they love diving deeply into their favorite books while under the herb's influence, while others (and especially newbies) may find that they get stuck on a single sentence and end up reading it over and over again. They may find they get stuck on a single sentence and end up reading it over and over again. And again.

# 6 SLIPPERY SLOPE OF PROHIBITIONISM

As noted in the history chapter, the war on marijuana is a very recent phenomenon that raises troubling questions: Why would a society ban its most useful natural resource and declare war upon a plant and lock up some of its most productive members who are not hurting anyone? Why would a government suppress science and promote bigoted lies to its citizenry? Why would Congress after Congress vote to fund a policy like the drug war that is universally recognized as a failure — up until the present day?

To answer the last question first, the drug war is only a failure if you believe its stated goals, to eliminate drugs and drug use. If you look at it from a big-government point of view, it is a huge success. Police budgets have received hundreds of billions of dollars, people of color rounded up and locked away, the assets of millions of people seized by the government and law enforcement for their own use, and trillions of dollars directed away from farmers, sustainable industries and herbal medicines and redirected to timber, mining, fossil fuels and pharmaceuticals. That is no accident.

## PHARMACISTS, BUREAUCRATS, RACISTS & INDUSTRIALISTS

The first wave of the war on cannabis came with the usurpation of medicine by global pharmaceutical companies, led by Henry Finger, claiming to "clean up" their industry by picking which patented drugs we can have and doctors can prescribe. The second wave came when Jim Crow-era racists began writing drug laws; first targeting the opium used by Chinese and later the cannabis used by Hindi, Muslims, blacks and Latinos. Then, along came Harry Anslinger, head of the Federal Bureau of Narcotics, who needed to find jobs for his G-Men — government agents who had been enforcing alcohol prohibition but were no longer needed. The Department of Agriculture, doctors and the hemp businesses stood in their way.

Anslinger hit upon the perfect marketing scheme: Demonize hemp under a slang word that was unfamiliar to English speaking people but sounded mysterious, foreign and possibly dangerous: "Marihuana." (It was even spelled wrong; the Spanish spelling is "marijuana."). Fabricate news accounts of crimes that did not happen and attribute all manner of mayhem to an innocent plant. Promote a combination of profane and subliminal messages through the popular media to cause panic and monger fear, then pop up with a solution to which there was no problem: Ban marihuana and criminalize its users. It was easy to manipulate the media with lurid tales that were never fact-checked, and crazed pulp fiction stories that did not even pretend to be true but were told as "warnings" to refine society.

Once big-pharma drugs, big government, big media and big business were all in bed together, the Marihuana Tax Act, prohibiting all uses of hemp or cannabis without a tax permit, slid through Congress before anyone knew what it was about. The bill's hearings were marred by repeated lies. Gardening became a federal crime.

## ASSASSIN OF YOUTH & *REEFER MADNESS* PROPAGANDA

Anslinger's most effective fiction was a magazine article he wrote called "Marihuana, Assassin of Youth" — a morality tale of a girl who is driven to suicide by the depravity of marihuana. It was made into the film *Reefer Madness*, which is still watched as a farcical parody of its own lies, but in its day, it was effective. It used subliminal messaging by juxtaposing images as certain words are repeated in order to create a subconscious link between the images of horror and depravity and the term marijuana.

Anslinger dreamed up melodramatic tales to scare the public. This is a scene from *Reefer Madness*, 1936.

That kind of forced linkage of fabricated fears to marijuana, no longer called cannabis or hemp by the media and government, was pounded into the American psyche for about 50 years. In the

1960s, this glamorization of the "killer weed" and outlaw culture led to an explosion of use and arrests.

There came a brief respite in the 1970s as backyard pot gardens blossomed and cannabis use became acculturated into every stratum of American society. States began to decriminalize small amounts for personal use. President Nixon considered hippie marijuana users to be enemies, so a new federal bureaucracy was created, the National Institute on Drug Abuse (NIDA). Its job: To research only bad things and nothing good about cannabis and assorted other drugs.

Nonetheless, it still looked like legalization was in the air — until President Reagan got elected. After passing conspiracy laws and property forfeiture, he militarized the drug war and established two new tools: urine testing to enforce job discrimination (and hair testing to target people of color in particular) and a multi-million dollar media campaign to stigmatize the cannabis community as lazy, egg-frying terrorists. He disowned all the centuries of knowledge of cannabis and said only NIDA-spawned research would be allowed; hence the DEA could permanently claim that marijuana was neither safe nor a medicine. Many Americans fell victim to mandatory minimum sentences ranging from five years to multiple life sentences.

That policy continued all the way through President Clinton and President Bush to today. The over-incarceration of people of color in the U.S. is a global scandal. We went from being the land of the free to the biggest prison state in the world, spurring construction of private prisons held by for-profit corporations. States cut their support for universities and schools to pay for them. The result: 20 million people arrested since 1965, escalating to a rate of 700,000 arrests per year. Under the George W. Bush Administration, more and more states began to assert their right to legalize medical marijuana and California legalized medical sales. Early on he seized a medical marijuana property from the city of West Hollywood, but he seemed to lose interest in the drug war over the years as it ran on autopilot.

The problem for the government is that cannabis saves people's lives and the planet needs it, too. When the best medicine for AIDS and cancer turned out to be cannabis and the best answer to many global problems turned out to be hemp, it shook the foundations of the prison industrial complex, but the federal budget has continued to fund blanket prohibition.

In short, prohibition means using a smokescreen of lies and deceit to incarcerate people, destroy families, steal money and assets, strip away rights and benefits, and force people to use alcohol, tobacco and pharmaceutical drugs instead of the safer alternative: cannabis. Its collateral damage of broken homes, lost jobs, personal income, drivers' licenses, organ transplants, student aid, reputation and even property has ravaged society.

## CEASEFIRE IN THE WAR ON KNOWLEDGE?

After the embarrassment of Nixon being forced to denounce his own commission, the federal government learned how to effectively stifle research into the therapeutic effects of cannabis. The DEA handed a federal licensing monopoly to a group at the University of Mississippi that has worked with NIDA for decades to restrict researchers' access and thwart attempts to find new treatments, even for severe problems like intractable epilepsy in children.

As a professor at the University of Arizona's College of Medicine, Dr. Sue Sisley labored for years to navigate the reams of red tape necessary to purchase research cannabis from NIDA, which under current federal law is the only legal source in the United States. The NIDA cannabis was intended for combat veterans who suffered from PTSD resistant to other therapies; intrigued by anecdotal reports that many veterans with this kind of PTSD had found relief in cannabis, Sisley persisted her way through the morass of federal bureaucracy, and was close to her goal of winning NIDA approval in 2014 — when the UA regents decided, without any warning or apparent cause, to fire her.

There are signs of a pending ceasefire in the war on knowledge; six months after Dr. Sisley's firing, her study was approved to go forward in partnership with the Multidisciplinary Association for Psychedelic Studies (MAPS). The Colorado Department of Public Health and Environment will pay to finance it.

Even Dr. Nora Volkow, NIDA's director, acknowledges the drawbacks of the federal chokehold on cannabis research. During a June 2015 Senate hearing on federal barriers to cannabis research, she endorsed the idea that her agency's monopoly contract ought to be amended. Dr. Douglas Throckmorton, deputy director of the Center for Drug Evaluation and Research within the Food and Drug Administration (FDA), agreed. "Yes," he told the committee, "I think there are advantages to a broad supply of varied marijuana."

## CONGRESS FINALLY ACTS AS GOP BASE SHIFTS

Soon, perhaps, this byzantine federal war on knowledge is coming to an end. U.S. Senators Rand Paul (R-KY), Cory Booker (D-NJ) and Kirsten Gillibrand (D-NY) made history when they introduced the Compassionate Access, Research Expansion and Respect States (CARERS) bill on March 10, 2015. As the middle part of the bill's name implies, CARERS would significantly expand access to research-grade cannabis in the U.S. and break down critical barriers to legitimate scientific inquiry into its use; within weeks, CARERS had been introduced in the U.S. House of Representatives with 16 co-sponsors, but has not yet been adopted.

The Democratic Party has provided most of the support for reform legislation, and is now being joined by a handful of Republicans to give some reform bills a margin of victory. One early leader in this new face of the movement is Dana Rohrabacher, a Republican from Southern California, who spent over a decade attempting to introduce an amendment to the federal government's annual budget that would eliminate all funding for the Department of Justice to interfere with state governments attempting to reform their cannabis policies.

After 10 years, Congress finally passed a federal medical marijuana bill in 2013 as the Rohrabacher-Farr amendment to the 2014 Continuing Appropriations Act. Fast forward to June 2015, when the U.S. House of Representatives approved seven funding amendments to divert money away from the Department of Justice; besides renewing Rohrabacher-Farr, the House eliminated DEA funding to interfere with state industrial hemp programs and earmarked $23 million from its budget to fund the testing of rape kits, child abuse prevention and the purchase of police body cameras. Banking reform is in the works. We don't know which, if any, of these policies will endure but it is clear that momentum is building.

On Capitol Hill, a narrative is gaining steam that sees some of the federal punitive measures of the drug war as being far too costly to taxpayers. In short, there is a growing consensus in Congress that its multi-billion dollar war on marijuana simply costs too much. One of the most decisive shifts has been the reframing of the "end of the drug war" argument as a conservative issue. It's just too expensive. Indeed, some of the top leaders of the next wave of reform are starting to be Republicans.

## STATE CANNABIS LAWS

**Legal Adult Use**
Colorado
Washington
Alaska
Oregon
Washington, D.C.

**Medical Legalization**
California
Washington
Oregon
Nevada
Arizona
Montana
Colorado
Massachusetts
Rhode Island
New Jersey
Maine
New Hampshire
Vermont
Connecticut
New York
Delaware
Maryland
Michigan
Illinois
New Mexico

**CBD-Only Legal**
Utah
Alabama
Texas
Florida
North Carolina
Minnesota
Nebraska
Oklahoma
Kentucky
Tennessee
Georgia

**Decriminalization**
Vermont
Massachusetts
Rhode Island
Connecticut
New York
Delaware
Maryland
Ohio
Mississippi

**Full Prohibition**
Idaho
Wyoming
North Dakota
South Dakota
Kansas
Iowa
Arkansas
Louisiana
Wisconsin
Indiana
South Carolina
Pennsylvania

**Industrial hemp**
California
Colorado
Delaware
Hawaii
Illinois
Indiana
Kentucky
Maine
Maryland
Michigan
Missouri
Montana
Nebraska
New York
North Dakota
Oregon
South Carolina
Tennessee
Utah
Vermont
Virginia
Washington
West Virginia

## CONSERVATIVE CALLS FOR REFORM

Increasingly Democrats, Independents and Republicans agree on one key bipartisan issue: the urgency of ending the federal war on cannabis.

Senator Rand Paul has championed industrial hemp legalization throughout his Senate career. "I want things to be decided more at a local basis, with more compassion," the senator and his co-sponsor of the CARERS Act said. "I think it would make us as Republicans different."

Ted Cruz

Rand Paul

Chris Christie

Credits: Gage Skidmore

After blasting President Obama's prosecutorial forbearance of state cannabis legalization as "imperialist" at a 2014 Texas Policy Foundation conference, Senator Ted Cruz reversed course at the 2015 Conservative Political Action Committee, calling legal cannabis states like Colorado "a great embodiment of what Supreme Court Justice Louis Brandeis called 'the laboratories of democracy'... if the citizens of Colorado want to go down that road, that's their prerogative. I personally don't agree with it, but that's their right." Other leading Republican figures are jumping on the bandwagon. Former Texas Gov. Rick Perry took the stage in 2014 to brand himself as "a staunch promoter of the Tenth Amendment" who also endorses the power of the states to set their own cannabis policies.

Republicans who break ranks from this position risk a stinging backlash. New Jersey Gov. Chris Christie, once labeled a frontrunner for the presidential nomination, alienated young Republicans (who support cannabis policy reform by a ratio of nearly 2 to 1) with a promise to use federal might to crush all state-legal marijuana, medical or not. And as of a June 2015 poll, a crushing 55 percent of Republican voters said that they would never, ever consider voting for Christie for president.

## THE CHANGING FACE & TONE OF PROHIBITIONISM

With more than half the voter population now favoring legal cannabis, prohibitionists have been forced to shift their arguments to accept some minor reforms. For example, modern prohibitionism advocates for allowing CBD medical marijuana only in the most limited form for the fewest people — and everyone else goes to jail. Industrial hemp for small research crops that are financially non-sustainable, but no vast fields of hemp to power industry, or you go to jail. More recently we hear that legal adult use means that big corporations get to control the market at an exorbitantly over-taxed price comparable or greater than the black market, and if you grow your own or sell at a better price — you guessed it — you go to jail.

The lies and contradictions continue unabated. During California's Prop 19 campaign, its opponents argued that the legalization plan was uncontrollable because it let every city and county set its own policies on sales and commercial activity, even allowing them to ban sales. When the legislature discussed regulating the sales of medical marijuana statewide, these same interests came back and argued that every city and county must have the power to prohibit both sales and personal cultivation to preserve "local control." Politicians, always looking for a compromise that gives the police whatever they want, appear to be willing to make restrictive "reforms," as long as they get to keep putting people in prison for marijuana.

## THE SHIFTING SAMS OF TIME

In the debates over the future of U.S. cannabis policy, "Just Say No" has ceded to "the third way." This new approach is best embodied by Smart Approaches to Marijuana (SAM), the leading anti-legalization group to emerge in recent years. Headed by Patrick Kennedy and Kevin Sabet, SAM sounds a far cry from the drug war rhetoric of hysteria from the 1980s and '90s. Arguing for a "third way" between legalization and prohibition, Sabet agrees with the approximately 70 percent of Americans who oppose any prison time for nonviolent cannabis offenses and support some form of medical marijuana model. Nevertheless, neither Sabet nor SAM endorse the idea of adult use legalization, despite the fact that a majority of Americans also support these reforms. So, the market will still be illegal, prison still will be the consequence, and drug tests will divide society: prohibitionism.

Patrick Kennedy

Kevin Sabet
Credits: Gage Skidmore

# 7 CANNABIS' NATURAL PLACE IN SOCIETY

People think they know how to spot a cannabis consumer: Cheech and Chong, the dude from *Fast Times at Ridgemont High*, hippies, slackers, dropouts, gangbangers, blacks, Mexicans, rock musicians, political protesters, their cousin who can't get or hold a job, and anyone wearing Birkenstocks. However, when their hardworking, middle-class next-door neighbor gets arrested for marijuana, people typically say, "We never even knew he smoked pot. He was such a nice person: It's so sad." Indeed it is sad, because, while stereotypes may contain some bit of truth in some cases, most are way off.

People from all walks of life in all parts of society use cannabis, including corporate presidents, professional athletes, students, teachers, artists and computer programmers. Only a small, targeted group ever got into trouble for it; most cannabis consumers go through life never getting arrested, just like anyone else. Take a look around you when you are out in the street. About 12 percent of the people passing by consume cannabis on an ongoing basis. A smaller number use it daily, and somewhere

**EVERY NEWBIE NEEDS TO KNOW:**

"Discrimination against Latinos, specifically racism aimed at Mexican-Americans, started the war on cannabis. We cannot afford to fear reform by continuing to buy into that prejudice. This is a huge opportunity for Latinos to repudiate the stigma of "marijuana," reclaim our legacy in connection with the healing cannabis plant, and create sustainable economic opportunities for our community."
— *Omar Figueroa, attorney at law*

## Traditional American Cannabis Subcultures

**Latinos** deserve first mention since they were the first group demonized with the word. A hundred years ago it was an alternative for workers to liquor but then lost popularity over the generations, due to harsh prison sentences and the enforced social stigma.

**Jazz** musicians were inspired by cannabis in the 1920s and celebrated it in many songs. "Satchmo" Louie Armstrong lamented that he had to give up "muggles" due to police harassment.

**Hippies** have been around for several generations now and are probably the best-known cannabis subculture. Their sharing traditions have been nestled in communities throughout the U.S., particularly in California and the Northwest. Other examples are Deadheads, Rainbow Gatherings, the Oregon Country Fair and Seattle Hempfest.

**Rastafarians** are the most clearly recognized cannabis religion by the U.S. courts. Known for their spliffs and dreadlocks, they can smoke huge ritual quantities.

**Hip Hop** culture often mixes a rap beat with bling, baggies and blunts. Its sometime espousal of violence and misogyny carries a lot of baggage.

**Patients** were the reluctant participants whose involvement with cannabis was for medical need.

**Growers and Dealers**. Yes, those guys. They've been essential.

between a third and half have tried cannabis at least once, depending on where you live. If you look at any group of at least 10 of your adult acquaintances, you are in all likelihood seeing at least one cannabis consumer and five or more people who have tried it.

If you wonder about the long-term effect of consuming cannabis, just look at the human race: we've been using it for millennia. So don't be surprised if you can't figure out who got high last night or last weekend, or who has never inhaled a single puff of marijuana or eaten a single brownie, and who's high right now. Society and culture are byproducts of cannabis interacting with humanity; many people simply are not willing to admit it. As the new paradigm of legal use and diminishing stigmas come into play, you can expect to see more and more consumers, but they will be harder and harder to spot. That's because anybody might be a cannabis consumer. Perhaps even you.

## THE CANNABIS COUNTERCULTURE

It is common in world history that, when a society feels threatened, its leaders and members turn against the "others" — people who look different, dress differently, come from foreign lands. Sowing prejudice and stigma, these elements paint the others as "deviant" and blame "them" for all problems. Society avoids the soul searching and responsibility taking for its own problems, finds a scapegoat and unifies the "us" part of society against the "them."

However, it is perhaps unprecedented in history that a government would target a common farm crop and garden herb with millennia of broad social acceptance, and then deliberately create a campaign of lies and bigotry to divide society from within. Curious indeed are the ways drug warriors have stigmatized their fellow citizens who continue in the global tradition of their forebears by growing and using that plant.

The initial trick was to fabricate lies and not let people know what plant they were talking about. Then police enforced the laws against racially identifiable groups within society, our black, Latino and Hindi Indian neighbors. Cannabis simply disappeared out of "White America's" medical kits, and when police told them they had to stop growing hemp, our farmers complied. When Prohibition ended, most whites stopped frequenting the jazz clubs singing about "that funny, funny Reefer Man," and went back to drinking. A handful of white jazz musicians like Mezz Mezzrow, and Hollywood stars like Robert Mitchum, would be busted to make the point but generally, as long as the Beat generation kept its profile low, it was people of color who bore the brunt of Reefer Madness.

This changed radically during the 1960s when marijuana use spread to the hippies, college students and the peace movements. When police beat down African-American civil rights leaders in the streets, it was one thing; when white people were beaten down beside them, that was another thing. So, when police began to beat down mostly white peace protesters in the streets, to raid white university dorms for marijuana and to send the hippie children of the white middle class to prison for years, the white news media shifted tone. By the time three of The Beatles and Mick Jagger of the Rolling Stones got arrested for marijuana, white faces blanched — not because they used marijuana, but that they would be busted for doing so.

Since the 1960s, cannabis has been acculturated and integrated into U.S. society to such an extent that its consumers are nearly invisible; in fact, tokers have been trained to hide in plain view. They grew it in their back yards in the 1970s and had to move the gardens into their garages in the 1980s. They survived by hiding it from their bosses, their children and most of their acquaintances. Some were finally able to announce themselves in the 1990s, as the medical legalization process unfolded and as long as they were patients. Then it turned out that there were more tokers than anyone had thought. Most still hide their identities, beat drug tests or start their own businesses. They live normal lives, sneak a puff here or chew an edible there without letting on. If they are medical users, cannabis may well be the only way they can live a normal life. As long as they work harder and drive more carefully than other people, nobody is the wiser.

During the new millennium they have shown up more and more in films and television shows — not solely as caricatures of a slacker but, first, as severely ill patients and their providers and, more recently, as nuanced characters who get together with friends and occasionally smoke a joint while they talk over life's concerns. Then, all sorts of hilarity ensues … or not. Pretty much like real life. Comedian and commentator Bill Maher has been the most out front and steadfast media champion of legalization. Hence, the authors contend that our cannabis consuming community is an acculturated American subculture that deserves to be honored, not criminalized.

## PRESIDENTS & CANNABIS: A TRADITION

*"I smoked but I did not inhale."*
*— President Bill Clinton*

*"I did inhale, frequently. That was the point."*
*— President Barack Obama*

## CANNABIS & MODERN SOCIETY

All across the country and every day, newbies are learning to see cannabis in new ways. People know it isn't the Reefer Madness demon weed, or even the "gateway drug" of the Just Say No era — but we want to know what we are getting into.

Health concerns related to smoking cannabis have been greatly overblown. Credit: Chmee2

How safe is cannabis? The truth is, for most adults, the use of cannabis is very safe; but there are a few caveats all newbies should know, just to be cautious.

When it comes to the general safety of cannabis consumption, let us refer to the writings of Judge Francis Young, who as an administrative law judge for the Drug Enforcement Administration spent over two years gathering the most comprehensive record on the effects of cannabis that had been compiled in nearly a century. At the end of these exhaustive hearings, Judge Young declared cannabis to be "one of the safest therapeutic substances known to man."

In the years since his pronouncement, the evidence for cannabis' safety has continued to mount, so that we can say with great confidence that it is a safe, nontoxic alternative to many legal drugs that may have nastier side effects. Nonetheless, there are a number of social issues to consider:

***How old is old enough for psychotropic cannabis?***

Everyone agrees that marijuana is not for kids. Most things in society get an age of consent for legal use; 16 for a driving learner's permit or to marry with parental consent; 18 to drive, sign contracts, join the armed forces, get married or buy tobacco; 21 for alcohol consumption. When it comes to cannabis, many people think 18 or even 16 years is old enough to use it, but most people opt for a 21-year age of consent — because of the alcohol standard.

While there is no solid scientific data showing significant risk to minors from THC, doctors typically prefer to approve CBD strains for pediatric use out of an abundance of caution. Many case studies show that

**EVERY NEWBIE NEEDS TO KNOW:**

"The most common side effects of marijuana are munchies and happiness." — *Kyle Kushman, cultivation author* children respond well to THC-based cannabis therapeutics and most people would agree that it should be allowed with a doctor's oversight. The human endocannabinoid system matures at around 15 years of age; in Holland the age of consent was 16 years for a long time and then bumped to 18 when social conservatives took control of the government — again a completely arbitrary number. Researchers now claim that the brain does not fully mature until age 25, but there is no data to suggest that anyone should wait that long to use cannabis; alcohol does actual, documented physical damage and drinking it is allowed at age 21. Cannabis is a neuroprotectant, not a hazard, so in reality it should be encouraged at a younger age than alcohol, which kills brain cells. The big concern with age 18 is that some people are still in high school, so the age of 19 has been widely recommended; after the endocannabinoid system matures and after graduating high school, it's still early enough to reduce alcohol consumption during college and hopefully beyond.

Ironically, marijuana use has become less glamorous to young people as the "forbidden fruit" effect has been blunted by medical use and adult legalization. Nonetheless, social convention in the U.S. suggests that 21 will ultimately be the accepted age for social use, as it is in four states as we write this book.

***Should grandma smoke weed?***

Some of the Americans who most vigorously oppose the use of marijuana are the ones who probably need it the most. If grandma suffers from arthritis, she may want to know that medical marijuana is one of the most effective, nontoxic treatments available. If she's concerned about euphoric effects, you should encourage her to try a topical cream or spray, which can provide quick and effective relief without crossing the blood-brain barrier, which is a fancy way of saying it won't get her high.

Should grandma smoke weed? Well, maybe not a blunt. Credit: Tibor Vegh

If she's concerned about the risk of Alzheimer's, Parkinson's or other neurodegenerative diseases as she ages, she should definitely use cannabis. Clint Werner, author of *Marijuana: Gateway to Health*, notes that researchers at the Scripps Institute found THC to have "remarkable inhibitory qualities" against the advancement of Alzheimer's disease compared to other available treatments. The *British Journal of Pharmacology* agrees, noting "cannabinoids offer a multi-faceted approach for the treatment of Alzheimer's disease by providing neuroprotection and reducing neuro-inflammation, whilst simultaneously supporting the brain's intrinsic repair mechanisms." Simply put — yes, your grandma really should use cannabis, but maybe not by smoking, and probably not by smoking big blunts.

***How dangerous is marijuana compared to other drugs?***

This chart by R.S. Gable compares cannabis to other common drugs in terms of its dependence and toxicity. The closer to the bottom left corner, the safer. Cannabis is the least toxic and less habit forming than coffee.

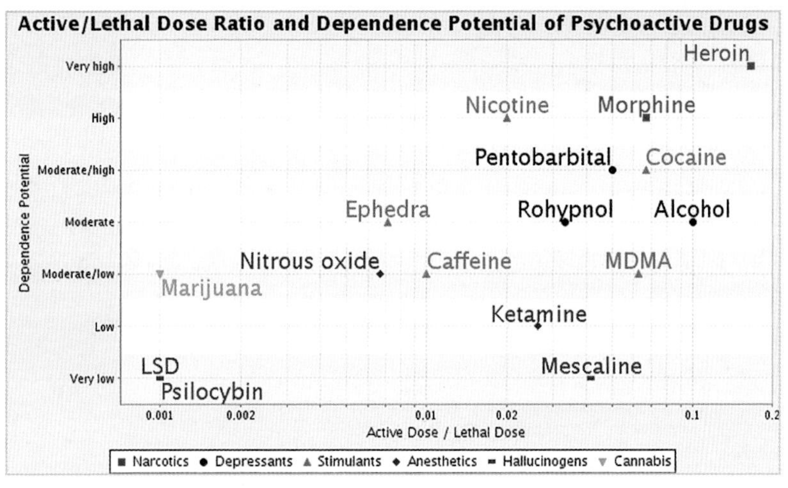


This chart compares the relative risks of various drugs. Lower left is safest, while upper right is the most dangerous.

***Driving and public safety***

Wherever people talk about cannabis, one question keeps coming up, time after time: how will legalization affect road safety?

While we would love to tell people that traffic accidents will plummet, unfortunately that's not the case. However, traffic accidents probably won't go up, either. The best available data, as of the time of this writing, is that the legal status of cannabis in a state doesn't seem to have any appreciable effect on the rate of traffic accidents in that state, one way or another.

So, if you feared the worst, you can rest easy. But even though we know that legalization coming to your state probably won't mean carmaggedon on the roads, we'd still hate to see any newbie become a statistic. So here are some basic newbie safety tips to make sure you're a responsible consumer who doesn't endanger the roads:

If you feel impaired, park the car for a while.

- Yes, cannabis use can impair driving. Fortunately, the effects appear to be mild compared to those of alcohol; cannabis-impaired drivers are more likely to drive more slowly, for example, than to drive too fast. At a minimum, wait at least 15 minutes for the inhaled effect to diminish, a half hour for tincture, and two hours for edibles. And always check your coordination and balance before you get behind the wheel. If you feel sleepy, don't drive.
- The effects of edibles come and go in waves. If you get a wave of strong effects, please pull over, stop and wait until the high wears off. Remember that it's impossible to fatally overdose on cannabis, but a car accident is another matter. That's not a risk worth taking, for a newbie or anybody else.
- Don't ever mix alcohol, cannabis and driving. Enough said. So, in other words, if you feel too high to drive, just park the car a while.

The other big question regarding society has to do with cannabis in the workplace.

## DRUG TESTING IS A FORM OF JOB DISCRIMINATION

Does marijuana use impair people's work or work ethic? The very existence of drug testing is evidence that it does not. If marijuana smokers exhibited low productivity or incompetence, they would be fired for that. But because they can't sort their toking employees out so easily, businesses rely on testing body fluids (urine, blood or saliva) and hair to identify the cannabis consumers on their work force, and then fire them. Drug testing was not accepted in business until the federal government began to require it, and, once that started, a new predatory industry was launched to prevent "pot heads" from having jobs. Hair testing is the most pernicious technology, since dark, textured hair retains drug residue longer than light, smooth hair, ensuring that people of color would be the first to lose their jobs. Coincidence? Not really.

> Drug testing adults for cannabis is a perverse practice that practices discrimination while letting truly impaired drivers stay on the road and incompetent employees keep their jobs — as long as they test negative. It is a neat metaphor for the drug war, imposed on society for unclear reasons and not meant to work but it lets some unsavory interests profiteer off the suffering of others. Impairment testing is the clear answer.

Which makes more sense: should you judge your employees by their productivity or pee? Workplace drug testing doesn't test for either impairment or job performance; it's ultimately a tool for job discrimination, and more and more drug testing is being applied to housing and education. Many sectors of the economy, such in creative arts and computer technology, refuse to test for cannabis because they need top-rate workers more than they need a cup of clean urine.

## PERSONAL, PROFESSIONAL OR ATHLETIC ACHIEVEMENT

Cannabis consumers are some of the most talented and productive people in society. Champion athletes are periodically stripped of their titles or suspended from playing over a urine test. Michael Phelps won more gold medals than any Olympian, and nearly had them stripped because he was caught toking a bong afterward. Former Beatle Paul McCartney has written more hit songs than anyone else, ever. All three of our most recent U.S. presidents have used it. Visit veryimportantpotheads.com and celebstoners.com to get a longer list of accomplished cannabis users; there are way too many to fit into this space.

## THE DOMESTIC U.S. CANNABIS COTTAGE INDUSTRY

When millions of adults went out looking for cannabis products in the 1960s, all that was available was imported from Mexico, where it was grown outdoors, harvested, dried and maybe cured, its seeded flowering branches pressed into bricks and wrapped for sale. It went for a steep price in its day: $200 per pound wholesale and $10 to $20 per ounce retail. If you paid $5 for a matchbox, that was pricey. A dollar a joint was the norm for many years, as well. Prices varied according to where you were located.

The business model was simple: someone would go to Oaxaca, Alcapulco or Michoacan in Mexico and drive home with a trunkload or a truckload of marijuana, smoke some, sell the rest to their friends and make enough money to drive back to Mexico for the next run. Their domestic counterparts opened so-called "head shops" where they sold pipes, books and other ancillary products while educating the community about cannabis.

When President Nixon launched Operation Intercept, the equation changed. You could still go to Mexico, but gangs began to take over the trade, leaving three options. You could work with the new criminal networks in Mexico, grow your own and buy locally, or head farther out to Colombia, Jamaica, Panama, Thailand and other places where the businessmen who ran the marijuana trade were friendlier. Maybe pick up a little cocaine or opium on the side. The more exotic approach was going to Afghanistan, Kathmandu, Turkey, India, Nepal and Morocco for sifted and pressed hashish.

Cannabis products flowed into the country by air and by boat. The quantity imported overland from Mexico kept prices low, brickweed being the norm and people working from home in the U.S. selling bags of seedy herb. Meanwhile, the domestic growers of Northern California were also bringing the next generation of producers into the cannabis trades, growing, trimming and selling; not making a lot of money but paying the bills. They kept the market pretty well satisfied until the mid to late 1980s. By then, better grade herb cost several dollars per joint and up to $40 per ounce retail in LA: Prices were higher, but not far out of line with inflation. The Dutch began to register and market seed lines, and *High Times* magazine launched its first cannabis judging competition.

Unfortunately, the drug war continued to spread like a cancer under President Reagan. Suddenly, U.S. cities were awash with cheap cocaine, and federal spending against marijuana spiraled out of control. Then when George H.W. Bush took office, he launched a foreign war on the relatively benign Mexican supply network, pushing it further into the hands of the most violent gangs who could stand up to the military raids, invading Panama and arresting his former ally who ran the country, General Manuel Noriega. He targeted Dutch seed suppliers for conspiracy to export to the U.S. He launched a domestic paramilitary war of aerial surveillance, herbicide spraying, thermal heat imaging of homes, shutting down grow stores and "head shops," and sent national guard troops marching through the prolific Emerald Triangle growing area in California. He terminated the federal IND program to any new participants and increased all the budgets yet again.

The so-called Prohibition tax went into effect, and the invisible hand of economic supply and demand went to work. Imported marijuana jumped from $40 to $100 an ounce almost overnight, joints hit $5 or more each, brownies fetched $5 to $8. The limited domestic supply moved quickly to gourmet sinsemilla grown in deep privacy and sold in the price range up to $500 per ounce, higher than the price of gold at the time. Growers were getting rich, trimmers were paid top dollar, big spending growers supported their communities, high risks meant high profits and marijuana became the nation's biggest cash crop. Before long, tunnels were being bored under the borders, officials in Mexico were being assassinated, and cartels moved through the global marijuana markets. More people got interested in the domestic supply side of the business.

President Clinton took office as the first semi-public dispensary opened in San Francisco. Cancer and AIDS patients were all welcome if they became members of the private club, at the discretion of its operators. Other underground outlets began to operate, as the domestic supply slowly began to catch up with the demand and the price of outdoor herb became stable in the range of $4,000 a pound in the West, and indoor weed still flirting with $5,000 a pound. CBD had been nearly bred out of the marijuana supply, and THC content of some strains spiked up to around 7 percent.

By the mid-1990s, The SF Buyers Club had grown to have 14,000 members, but was raided and shut down in the midst of the 1996 Proposition 215 medical marijuana campaign in California, reopened soon after the law changed, and then was shut down again when the state Supreme Court ruled against its operator, gay rights activist Dennis Peron. Around San Francisco and across the bay in Oakland, new groups of people came into the compassion business to fill the gap. Commercial operators began to expand operations into residences, back yards and small, discrete warehouses. Medical marijuana advocates were getting better at playing the political game and backing candidates and initiatives. The domestic grows were getting larger as the century came to a close but remained small compared to the hectares of fields growing in Mexico.

## THE LARGE-SCALE CANNABIS INDUSTRY EMERGES

Once the medical-use market opened up in the new millennium, people began to talk about a "Green Rush" that embodied both green herb and even greener cash sales, particularly in California, where tolerated or illegal dispensaries began springing up around the state to serve the needs of paying medical patients. The legislature's SB420 legalized sales among patients hoping to gain control of the business end of the access equation but, because of the federal prohibition against banks, it remained a cash business. A new generation of cannabis entrepreneurs emerged. Efforts to regulate were thwarted by federal raids, but dispensaries kept closing and opening. California demanded its sales tax, legal or not, and dispensaries complied and implemented a "good neighbor" policy of reducing nuisance and improving the neighborhoods where they located. Oakland and other cities began to demand permit fees and zoning compliance, and the statewide system of individual patients growing for patient-run dispensaries grew into an ever-larger market. That market was interested in smoking, edibles, tinctures, beverages, concentrates, vaping paraphernalia, grow books, grow equipment, garden supplies, seeds, clones, processing equipment, real estate, office and retail locations, attorneys, accountants, licensed contractors and so on. So, ancillary businesses took root as specialty websites, newspapers and magazines began to track and promote marijuana outlets, strains and products.

The fact that it has been relatively easy for patients to become qualified for the state's supply network ensured that very skilled growers would become involved and they would have a large and expansive market. The ease of entry into the producers' market, the big spike in demand and the difficulty of law enforcement to secure convictions for cultivation under the collective system caused a big jump in production. The surge in supply brought THC content up to around 10 percent and the wholesale pound price down to $2,500 for outdoor bud and $3,500 for indoor, and the growers responded with yet another large surge in production, powered by larger locations and by drastically improved growing techniques used by a greater number of growers. People continued experimenting with genetics, recipes and growing techniques in a flurry of creativity. Oaksterdam University opened to train people to move from the counterculture to the over-the-counter and how to bring good business sense to the cannabis trades.

In the second term of the G.W. Bush administration and heading into the Obama years, two new developments shook the cannabis markets. Groups with alleged ties to drug gangs in Mexico began making massive outdoor grows in U.S. state and federal parks, and growers stopped dumping out their shake and trim and began to turn trash into stash. The first trend was short lived due to massive aerial law enforcement efforts and their sheer enormity of size. The second trend fueled an explosion in non-inhaled product availability such as edibles, topicals and tinctures, and also in the supply of concentrates like water hash, bubble hash and ice hash that brought medical cannabis to the next level. Later, many solvent extract producers came to rely on the shake supply for their raw material.

## RE-LEARNING THE VALUE OF DISCRETION

President Obama's ascent to office was a moment of great joy in the cannabis industry. THC content in bud sometimes surpassed 20 percent. Hundreds of thousands of licensed and unlicensed dispensaries and delivery services operated in California. Commercial grows were getting local approval and growing larger and larger, into the dozens, scores and then a hundred or more lamps at a location. It had begun to affect the electrical grid and people were willing to be transparent and out in the open in the expectation that the administration was benevolent or even friendly to them. They learned to their dismay in 2010 that telling police you were complying with state law would be taken as a confession that you were breaking federal law. Text messages, emails, corporate documents, state filings, taxes — all were used against the providers. Raids wiped out the legitimate Montana medical cannabis supply network, invaded California and dealt a big setback to the process of bringing the cannabis industry above board.

A lot of growers went back underground after the crackdown on California outlets, and the DEA soon reported higher rates of diversion out of state or into the unregulated illicit market. While the feds waged war on Montana and California, Colorado rose to the top with its Amendment 64 legalization for non-medical use, and Washington passed its deeply flawed Initiative 502. Both relied on heavy taxation and burdensome bureaucracies to put a drag on the market, but Colorado's rollout has been so positive that it helped Alaska and Oregon in the passage of their initiatives in 2014. Meanwhile, a new generation of warehouse grows and high-yield growers have pushed the wholesale prices of marijuana in California down to record lows, in some cases around $1,000 per pound for outdoor weed and as low as $2,000 per pound for indoor bud. But yet, the price to the end consumer is not so different than it has been for all these decades.

Discretion remains a core principle of the cannabis industry. Even today when police raid large-scale gardens in commercial areas and whole houses

## Comparing Tobacco With Cannabis

"The culture (of tobacco) is pernicious. This plant greatly exhausts the soil. Of course, it requires much manure, therefore other productions are deprived of manure, yielding no nourishment for cattle, there is no return for the manure expended.... It is impolitic. The fact well established in the system of agriculture is that the best hemp and the best tobacco grow on the same kind of soil. The former article is of first necessity to the commerce and marine, in other words to the wealth and protection of the country. The latter, never useful and sometimes pernicious, derives its estimation from caprice, and its value from the taxes to which it was formerly exposed. The preference to be given will result from a comparison of them: Hemp employs in its rudest state more labor than tobacco, but being a material for manufactures of various sorts, becomes afterwards the means of support to numbers of people, hence it is to be preferred in a populous country. America imports hemp and will continue to do so, and also sundry articles made of hemp, such as cordage, sail cloth, drilling linen and stockings." — *Thomas Jefferson, Farm Journal, March 16, 1791.*

growing cannabis in residential neighborhoods, the next-door neighbor's refrain to the news camera remains the same: "I had no idea they were growing/selling marijuana over there. They seem like such nice people."

The dichotomy today distinguishes between the needs of the cottage industry, those of the mass industry and those of the regulatory bureaucracy, who see cannabis as a cash cow to fund various other concerns. The so-called Prohibition tax has been replaced by all sorts of tax levies, and the risk is that when illicit market prices for cannabis go down and regulated prices go up, the policy tends to prop up the very market it seeks to supplant.

Green Flower Media is a new group focused on educating people about the facts, science and benefits of cannabis, and ending the social stigma of cannabis consumers through "coming out" videos. It provides downloadable reports on such topics as "how to talk to your kids about cannabis" and "the safest and healthiest ways to consume cannabis today," and more. Check it out at greenflowermedia.com.

# 8 SINSE AND SENSIBILITY

Congratulations, newbie — you're about to graduate to cannapreneur. For you, merely appreciating cannabis isn't enough; you want to get into the industry, too.

It's time to assess why you are getting into the trade, find your entry point, go through the steps to get there, find and fill a role that you can maintain and hold over an extended career path. Then consider your exit strategy for when it is time to sell off, retire, release your energies or move on to another position.

Resin is crystal looking to the eye, fragrant to the nose, flavorful to the palate, sticky to the touch.
Credit: Garrettaggs55

Because each situation is as unique as you are, let's start with the questions you need to consider as you seek your own answers. What is your driving motive? Is it because you heard this is the new "Green Rush" and want to make a lot of money? Is it because you are altruistic and want to help people who are sick and need medical relief? Is it because you love this plant and want the personal fulfillment of working in a career field that lets you connect with cannabis and the cannabis community? Do you love to work with the public and to extol the wonders of cannabis? Is it because you like to design and promote gizmos and have a lot of ideas you want to bring to market? Do you love science? Did your vegetable garden and botany classes destine you to improve upon the basic plant or gardening practices? Do you love to bake? Are you concerned that the industry could end up in the hands of devious corporations and you want to preserve the community ethic? Is it because the local grocery store doesn't have any job openings? Are you merely curious about this new and emerging industry and want to see what it can do for you and you can do for it?

People who are motivated primarily by money may want to look at the investment potential of startup businesses and evaluate their options.

**EVERY INDUSTRY NEWBIE NEEDS TO KNOW:**

"While marijuana legalization has majority voter support and more states are changing their laws, the movement that is pushing for these reforms still has a long way to go. Federal law remains unchanged — for now — and the next presidential administration could work to reverse our gains. Therefore, it is important that anyone who wants to be involved in the marijuana industry — whether as an investor, business manager, sales representative or even just as a loyal customer — absolutely needs to prioritize supporting the movement and the nonprofit organizations that make the emerging legal industry possible in the first place. If we don't band together and flex our political might, the days of having to buy marijuana from the guy on the corner instead of from a legitimate business could return much sooner than most people think."
*— Tom Angell, Chairman and founder, Marijuana Majority*

Record keeping in the cannabis industry may require reams of paperwork. Credit: Isaac Bowen

They may want to maximize profits, but to do that means maintaining the quality of their products and the value of their business as they look ahead to selling out to bigger investors down the road. Taking too many shortcuts can cause big problems; being careless can end you up behind bars. Many people have lost money in the green rush, and scammers are ready to take advantage of ill-advised funders. Watch your assets.

People who are interested in the medical value of cannabis need to understand that not every medicine works equally well for every person, and that the means of ingestion has a big impact on its effect. Will you be working with cancer patients, with senior citizens, in a hospice or cannabis health resort? Are you mentally and physically strong enough to work with people in constant pain, lacking the ability to control their behaviors and bodily functions? That work takes special dedication, so be sure you are cut out for it. Knowing how to pair the right cannabis products to alleviate specific symptoms is a valuable skill, whether you are a hospice worker, budtender, an advisor, a medical professional or clinic support staff. If you simply love being around processed cannabis, maybe you would like to be a grower who specializes in producing top-grade flowers. Budtenders savor the aesthetics

Extracting supercritical CO2 resin wax. Credit: Wiki: seics

of cannabis, judge the various strains by appearance and fragrance and inspect for contaminants to gauge the quality of buds and extracts. They also help others learn to appreciate those qualities. If you love working with the public and talking to people, this is a great job to consider. It can open the door to other retail and management positions, and anybody who comes in the door is a potential contact for some later enterprise.

Vape pen

If you have a scientific bent, you might want to pursue a career in medical research or work in labs that handle product testing and quality control. If you like to invent and market consumer devices, then consider the ancillary industries — designing, prototyping, manufacturing and marketing packaging, pipes, vape pens, garden equipment and supplies. This is a very competitive area of the market, so make sure nobody else invented your gadget first.

If you love working with plants, do you also have the physical wherewithal to do the heavy labor? Do you have allergies to working with raw plants? Growing involves knowing how to provide a consistently high-quality crop that is free from contaminants, being willing to experiment with new techniques to improve what you already have going, and being able to cope with unexpected and potentially catastrophic garden adversity. Outdoors, you have to contend with nature, and indoors, with technical problems and equipment failures. Are you up for a challenge?

If you love to bake, the edibles market is an option. Remember, however, there is a big difference between cooking for friends and family and working in a high-production commercial kitchen facing deadlines and liability issues.

**EVERY INDUSTRY NEWBIE NEEDS TO KNOW:**

If you make edibles for the cannabis market, remember that some members of the community have food allergies and other sensitivities. Some decline to eat meat, and some have low-sugar or low-fat diets. How will you cater to them?

**EVERY INDUSTRY NEWBIE**
**NEEDS TO KNOW:**

"If you plan on opening a dispensary, make it professorial and medical. Hire a nurse or pharmacist to manage it and hire mature budtenders. Have your staff wear identifying shirts with logos. Have accurate information about your strains. Test and label each of your products." — *David Bearman, M.D., Physician, Expert Witness*

## PROTECTING THE LONG-TERM RETURN ON INVESTMENT

Doing things right and acting ethically are keys to protecting the future of the industry. If you are interested in preserving and protecting the ethics of the community and the small players who have built up this industry to its current and future potential, ancillary services could be a great place for you. It could be as a consultant, accountant or legal advisor; as an investment advisor; a lobbyist working on legislation; a mediator to resolve disputes; or an attorney to protect the interests of inventors, investors and consumers. It could be a civil job inspecting to make sure kitchens are clean, gardens are organic, locations are up to code and consumables are safe with clear labeling, non-wasteful packaging and measured dosages.

If it is simply because you don't like your other job options or are curious about what is entailed in moving forward in this career direction, be aware that this is a crucial time in the development of this industry and it carries real risks. It would be a thrill to be the test case that decides whether the federal death penalty really will be applied to you as a marijuana grower or seller — but maybe that's not the kind of thrill you want. We are moving from the illicit market to a regulated one, but along a murky legal path through federal prohibition, relaxing federal policy, tightening state

No matter what your motive or which opportunities you intend to pursue, be advised that there are a lot of other people who have their eye on the same goal as you. Getting ahead of the cannabis curve by doing research with the *Newbie's Guide* is a good strategy, but also consider attending an institution like Oaksterdam University to give you hands-on exposure and practice working with cannabis and allow you to hear from and work with top experts in the field. The more prepared you are going into the industry, the more pleasure, satisfaction, fulfillment and financial rewards you can get from your career.

Oaksterdam professor Chris Conrad wears a hemp shirt in a medical marijuana garden.
Photo by Mikki Norris

regulation, local zoning and NIMBY (not in my backyard) neighbors that can lead to perplexing, sometimes insurmountable obstacles.

Some conflicts are to be expected, but they don't have to be stoked up or made worse by inflammatory rhetoric, so choose words wisely, use them carefully and approach people and markets with empathy. Because of the stigma cannabis has borne for many generations now, there is a higher standard of ethical conduct required. Your challenge will be to maintain high ethical principles as you achieve your financial, career and life goals. Like women and minority groups, cannabis aficionados have to work twice as hard to get half the acceptance. Someday that will subside, but that day has not yet come. In the meantime, evaluate your current circumstances and keep up to date on the trends that can determine your future.

Nothing is simple in the cannabusinesses. The industry comes with a political movement and a community attached. The cannabis community is engaged with the industry and entangled with the drug war industrial complex, and the political interests of these groups intersect at certain points and diverge at others.

You will find some of them are the nicest people you've ever met — but not all are, so rely on sound judgment, not blind trust. Be ready to engage both your common sense and your physical senses in this task and to get close to the plant at the heart of your interests and activities. This is true of the industrial hemp products as well as the cannabis flower product markets. We suggest that you consume the flower at some point if you want to be part of the industry.

## EVERY NEWBIE NEEDS TO KNOW:

When you obtain cannabis, be sure you know that it has been laboratory tested for potency (i.e. specific cannabinoids, terpenes and terpenoids) and for quality assurance (the cannabis is free of mold, bacteria, fungus and pesticides). This applies to flowers, concentrates and edibles.
*— David Goldman*

Use all of your senses. Be observant and use your ears to listen to people and hear what they have to say about your products and processes. Don't take up every idea that comes your way because when you are invested in an idea, people's criticism can seem cruel and dismissive. Don't take it personally or ignore it; hear it as if spoken in the spirit of trying to improve your product or guide your process. When someone says they don't like it or won't buy it, ask more questions to find out why. If it's a legitimate reason, take heed.

At its core, a cannabis business is still a business. Be clearsighted: don't ignore your common sense. Where do your personal, financial or political interests lie? Is this a product or service with a market? How does it tie into other markets?

For example, if you design, manufacture and market ancillary products, pay attention to the aesthetic and functional aspects of your design. Will they generate demand? To have an ongoing market you need to have an expanding shopper base, a consumable, or a growing list of products to sustain your income. A single product that is offered in a saturated market or a fad that has outlived its era is not a tenable basis for long-term or expansive income.

## IS THIS THE RIGHT LOCALE FOR YOUR BUSINESS?

There are many other examples of where your senses and common sense come into play with the cannabis industry, but let's just look at one more. Say you plan to open a cannabusiness and you think you have a good location. There are a several considerations that require sensible planning and compromise such as price, proximity and security. Get your senses involved as well. Before signing a lease, walk the neighborhood. Is it by a school, a daycare or a park? Cannabis is not a danger to children; that's not the problem. It's the still-entrenched prejudice against marijuana that can create the problems for you. Parents have been conditioned to be afraid of your business and its clientele, and that is probably not a fight you want to take on as you launch an enterprise.

So look around and listen: Are there a lot of kids playing in the neighborhood? Close to a rehab center? Those are probably not good indicators for the location. Talk to people and listen to what they have to say. Ask around. Is there a local church with a sanctimonious anti-flower agenda? Is the

Not a good place to set up a cannabis retail business nearby. Credit: Mk2010

neighborhood plagued by burglars? Both of these are red flags. To be a good neighbor you must know your neighbors, so learn about them. If you plan to run a health resort or a cannabis-friendly hospice, you might want to be in an area that is convenient to health and support services. Once you have a business going, take regular walks outside to take a look around and smell the ambient air. If you can detect the odor of growing marijuana, so can your neighbors and so can thieves or police who are out to cause you problems. If you are in a more cannabis-friendly zone, however, that fragrance could attract you new customers. All these considerations need to be factored in to your process. But enough philosophy for now — let's get down to brass tacks in the next chapter.

**EVERY INDUSTRY NEWBIE NEEDS TO KNOW:**

Carbon filters can scrub exhaust of the smell of cannabis.

# 9 CONNECTING WITH YOUR STAKEHOLDERS

Now that you think you know almost everything there is about cannabis, it's time to open your business doors and find customers — or is it?

Not quite. There are two steps remaining. One is to work with a qualified attorney and accountant to make sure your business plan complies with all state and local laws (see Appendices) and appears financially viable. The other is to expand your mind a bit.

No, not like that. We mean that any successful cannapreneur must contend with an aspect of this particular industry that is more complex than it is for others: The community of stakeholders surrounding a cannabis business encompasses a wider and more diverse group of people than most other businesses.

In the world of business, directors look out for shareholders; cannapreneurs look out for their stakeholders.

**EVERY INDUSTRY NEWBIE NEEDS TO KNOW:**

"Traditional business models and traditional marijuana models don't apply here. It's important to forge a partnership between people with business experience and people with cannabis experience because the world of cannabusiness exists right in between the two and you have to be ready to interact with either one."
*— Alex Rogers, Executive Producer, ICBC*

Anecdotally, most of the problems of doing business in the cannabis industry arise out of the relationship the cannapreneur cultivates — or fails to cultivate — with neighbors. Take care of their concerns promptly or the police may decide to take care of them for you!

**LEGAL ISSUES**

Depending on the state in which you operate, you may be required to organize your business in some form of collective ownership structure, like a cooperative. Make sure you know your state's laws, because they have a profound effect on how you conduct business!

What do we mean by stakeholders? Unlike in gambling, a business' stakeholders includes its shareholders — those who directly own and control the company — and anyone else who feels they have an interest — a *stake* — in the outcome of the business' success or failure, including directors, employees, customers and neighbors. The idea has gained some traction in the business world in general, as scholars and businesspeople find the limits of focusing exclusively on quarterly profits and turn instead to the "triple bottom line" of people, planet and profits. A broad stakeholder strategy is a good idea for any business, but for a cannabis business, it is essential.

Let's explore the various kinds of stakeholders that a cannapreneur may have to address in order to succeed:

- Your partners, shareholders, suppliers, etc., all have a stake, naturally.
- Your clientele have to be happy enough to become return customers.
- Your neighbors can be just as important to its success as its patrons, as any collective that generates antagonism from a critical mass of neighboring families and businesses may soon find itself run out of town. Any sustainable facility must have a plan for engaging its neighbors to ensure that they feel heard and are comfortable.
- Your local government will limit your decisions via its licensing, regulatory and zoning processes. The planning department, fire inspectors, police and — you guessed right — numerous tax collectors all feel that they have a vested interest in the way your

business operates. Any collective planning to stay in business for very long has to find a way to work with these civic institutions. Remember: fair or not (most probably not), they got here first.

- Your employees are vital to success, especially if the owners and operators don't plan on doing all the work themselves. All businesses — but especially one in the cannabis industry — must find a way to keep its employees satisfied or risk catastrophic ruin. One disgruntled employee can cause enormous damage to an entire company. Under most state laws, the owners of a cannabis business are held responsible for the actions of their employees, regardless of whether they condone them.

Take a moment to review your business plan and consider how many different kinds of stakeholders will be affected by your decisions. Reaching out to them and hearing their potential concerns should be one of your earliest priorities.

A cannapreneur needs to know this about stakeholders: there are two fundamentally incompatible ways to perceive this plant. One perspective holds that cannabis is a drug of intoxication and abuse; the other that it is a harmless medicine. For example, one neighbor stakeholder may believe that cannabis is a vice that must be hidden or suppressed, even if they admit that legal drugs like alcohol are worse. Conversely, a patient may insist that medical marijuana is the panacea they need to smoke all the time, anywhere they want and everybody else better get used to it because it's their right. Those deeply held beliefs are based on nearly opposite perspectives.

Patients must fight for local access. Credit: Shay Sowden

"[T]he risk of marijuana use in states before passing medical marijuana laws did not differ significantly from the risk after medical marijuana laws were passed. Results were generally robust across sensitivity analyses, including redefining marijuana use as any use in the previous year or frequency of use, and re-analysing medical marijuana laws for delayed effects or for variation in provisions for dispensaries." — "Medical marijuana laws and adolescent marijuana use in the USA from 1991 to 2014: results from annual, repeated cross-sectional surveys," *Lancet Psychiatry* 2015: 2:601-08

Public opinion on the question of whether cannabis is a vice or a medicine has shifted sharply in our favor over the past two decades, and it is seen as a non-issue that might as well be legal. To engage your foremost group of stakeholders, the general public, you must see both sides of the issue — cannabis as both a vice and as a medicine — and find a path through that minefield while taking fire from law enforcement, biased neighbors and officials. If you get it wrong, it could be a public relations nightmare and bankrupt your business.

It's useful to force yourself to swap perspectives and take on a view antagonistic to your core business. Put yourself in the shoes of someone who only sees cannabis as a drug of abuse, and imagine what your business looks like from *their* perspective. A fraught mother with two children ages 13 and 16: How will *she* react to the news that you're opening a dispensary in *her* neighborhood? What are her fears? How can you address and alleviate those fears?

A police officer lives or works in the neighborhood with a 20-year career on the drug war's front lines. How can *you* break through to him?

An elderly woman lives nearby. She doesn't think nice people should go to jail for pot, but she gets alarmed at scary news claims that adolescent cannabis use might trigger schizophrenia! It's not true[1], but she's afraid of the possibility.

There is ample evidence to rationally disprove their fears, but facts alone rarely change people's minds. When was the last time *you* changed your mind about a deeply held belief after encountering a new piece of data? That's probably more recently than have most people. So what, in the hearts, minds and shoes of your most vocal opponents, will bring them to support your business?

## BUSINESS FOUNDER, MOVEMENT AMBASSADOR

To succeed with antagonistic stakeholders, you will have to go above and beyond the debate; you need to become an ambassador for the movement. Don't go into your community to fight — go in to win people over.

What's the difference? Whereas arguments are fought between adversaries, ambassadors seek mutual solutions through diplomacy. This is human social psychology at work: a person who comes at us with an argumentative tone tossing out facts and figures will probably be tuned out immediately. A person who signals that she understands us, wants us to be happy and seeks mutual solutions will quickly ingratiate herself into our favor, and we want to help her, too.

Showering the nervous mother with scholarly text documenting that states with medical marijuana laws have no increase in teen use (see sidebar on next page) probably won't get you very far. Instead, call a town hall meeting to discuss solutions that avoid teen use. Give all concerned residents an opportunity to voice their thoughts, and genuinely listen to them. Then *prove* that you heard everyone by being prepared to adapt your plans to accommodate them, like added I.D. checks and security. This is far more effective than simply laying out your case and expecting them to accept you on your own terms.

Remember, the police officer has a difficult job and sees society at its worst, so cannabis reform may feel like a major defeat for him in a thankless war, or like a waste of his time and resources. How can you frame your enterprise as a mutual victory? Expecting their dispensary in Oakland,

Successful cannapreneurs learn how to listen.

California, to draw the heat, Harborside Health Center founders Steve and Andrew DeAngelo planned from the beginning to partner with law enforcement. So when they hired staff for their dispensary, they also hired security and offered to help Oakland police by patrolling a radius of four blocks in every direction from their facility. Nearby residents began reporting that they felt much safer, and Harborside's relationship with OPD took root to the point that when Harborside got hit with a civil asset forfeiture action by the federal Department of Justice, the City of Oakland sued to intervene on their behalf, noting the loss to public safety if it lost Harborside. When your city sues on your behalf and not against you, you've succeeded spectacularly as a cannabis ambassador.

**EVERY NEWBIE NEEDS TO KNOW:**

Contrary to the fears of mothers nationwide, the legalization of cannabis for adults has not led to increases in teen use — in fact, if anything, reform seems to have the opposite effect. That, at least, is the conclusion of the National Institutes on Drug Abuse (NIDA) in their 2014 survey, Monitoring the Future, which found that while various states legalized cannabis, teen use actually went down.

Finally, that elderly woman nervous about the arrival of a pot shop over her fear of the unknown could turn out to be your best ally once you introduce her to topical cannabis medications that soothe arthritis, stiff joints and other daily aches and pains. Tell her about CBD oil not having psychoactive effects. She always heard marijuana was a smoked drug to get high, so when she discovers that your store will be selling traditional medicines like cannabis creams, tinctures, etc., this circumvents her deeply ingrained ideas to present her with a new perspective regarding cannabis; not as a bad habit but as a good medicine.

Come to the cannabis industry to win, not just to fight.

"As a retired police officer I believe that everyone should know that our government has misled the public about the efficacy and safety of cannabis. By bringing the cannabis industry above ground, it will protect patients and consumers while helping us to end a policy based on ideology, not on science."

*— Diane Goldstein (Police Lieutenant Retired), Executive Board Member, Law Enforcement Against Prohibition*

## SOCIAL BENEFICIARIES

**EVERY INDUSTRY NEWBIE NEEDS TO KNOW:**

"Presentation is everything. Always look professional. In this industry, an ounce of prevention is well worth a pound of cure."
— *Scot Candell, Attorney at Law*

You don't have to like drugs to loathe the drug war. The war on cannabis affects us all, whether or not we use it. It has a chilling effect on medical research, ships homegrown jobs overseas, and trumps the liberty that inspired the country's founders. Cannapreneurs have the power to reverse the flow. As we succeed, public costs can become public gains. As our businesses grow, so grows the domestic workforce. As more consumers turn to the regulated legal market, fewer will give money to the organized crime networks. As we keep our patients' supply safe, surrounding neighborhoods become safer. As our sales grow, so too do local tax receipts for our schools, fire departments and infrastructure.

## PUBLIC ENTITIES

Across the U.S., the infrastructure investments of previous generations are crumbling. In our roads, rails and bridges, the decay is literal; in public education, it is that and figurative. The cause is clear enough: for decades, the largest corporations have left America's shores for cheaper labor and lower taxes abroad, siphoning the country's wealth and decimating the workforce.

We have an historic chance to reverse this trend. No one likes paying taxes. But for cannapreneurs at the delicate first stages of the industry, the benefits they accrue from being good taxpayers are substantial. The American cannabis industry, like the plants which supply it, is growing rapidly at home.

Taxes fund the infrastructure that we all rely upon. Credit: Oregon DOT

In Oakland, which like many U.S. cities has struggled in recent years to balance its budget, Harborside did not fight the city's license fees and sales tax on medical marijuana but decided to pay generously to the local coffers. They also gave hiring preference to staffers who live in the neighborhood where they set up, to help stimulate the local economy. The strategy paid off. Oakland intervened in a federal suit against Harborside and asked the judge to keep the collective open and paying taxes. The dispensary remains open as we write this.

## SOURCING YOUR SUPPLY: SUPPORT LOCAL BUSINESSES

Business is booming. The local community loves your products; you can hardly keep them on the shelves. Where do you go to re-stock? You sometimes can get business supplies cheaper over the Internet or at a big box chain than at the mom-and-pop store around the corner, but the smarter move is to invest in the goodwill and economic development in your local community.

When cannapreneurs source their products or use their earnings to support other local businesses, the whole community benefits and your local customer base has more cash to spend. Everyone wins. Being part of a business association or Chamber of Commerce will pay dividends on your dues money.

## JOB OPPORTUNITIES FOR DRUG WAR VICTIMS

Few drug war POWs are tattooed, gun-toting thugs. Most of those whose lives have been shattered[2] by the drug warriors are nice people who don't fit the criminal stereotypes at all. They have families, hopes, dreams and aspirations.

They need our help. Practically every job application in the U.S. has a check box that most will simply gloss over with nary a thought — where we are required to disclose, under penalty of perjury, if we have been convicted of a crime. Sure, a bank would not want to hire a convicted bank robber, but marijuana "offenders" are different.

The drug war makes "criminals" of far too many people who should never have been arrested in the first place, and to bar drug war POWs from legitimate employment on top of the time they have already served

While we recommend paying local taxes as a token of genuine goodwill, federal taxes are another matter entirely. When filing taxes with the IRS, make sure you work with a CPA or attorney who understands the special rules for cannabis businesses, like section 280E.

is not only senseless cruelty but also political insanity — cut off from job opportunities, many have little choice but to return to the activities that got them in trouble in the first place.

It doesn't have to be this way. Cannapreneurs in the legal industry can help alleviate a serious problem with marijuana recidivism and gain skilled employees as long as regulations don't ban hiring ex-cons. Non-violent drug war POWs may even be better workers. Their history of arrest and conviction may tend to make them more responsible, since no one is more acquainted with the consequences of failure or will be more mindful of not messing up in ways that could jeopardize your business — or jeopardize their personal freedom.

**EVERY NEWBIE NEEDS TO KNOW:**

"The business of cannabis is a civil rights issue first, and a business second." — *Richard Lee, founder of Oaksterdam University*

**EVERY NEWBIE NEEDS TO KNOW:**

"The cannabis industry isn't just about making money, it's about repairing the damage done by the war on drugs."
— *Amanda Reiman, PhD MSW I Manager, Marijuana Law and Policy Drug Policy Alliance*

L.A.-based Homeboy Industries is one powerful example. This retail collective takes the hiring of drug war POWs into overdrive. Not only does it hire ex-convicts, they hire *only* ex-cons, and former gang members in particular. They offer each employee the chance to earn an ownership stake in the company and a seat on the board to vote on the company's future direction. Now the company owns multiple restaurants and clothing stores and must constantly expand just to keep up with its own success. It is vivid proof that, given the choice, countless ex-cons would rather choose to have legitimate careers than return to the gang.

Likewise, the legal cannabis industry should be a model for vetting and hiring non-violent drug war POWs and advocate for changes in the laws that now risk government-codifying backward thinking as the

Non-violent drug offenders can be trustworthy, hardworking employees if given a chance, as Homeboy Industries has shown.

legal precedent. Legalization laws like that in Washington State explicitly discriminate against ex-cons in the industry. The state's I-502 regime forbids anyone with any criminal record from holding a license to grow or distribute cannabis. While it doesn't forbid license holders from hiring cannabis POWs as employees, it does unfairly deny such people from being company owners.

It behooves all cannapreneurs who have been fortunate enough to avoid the criminal justice system to act in accordance with human compassion and their own financial self-interest; give the victims of the drug war a second chance. Under different circumstances, you might have been in their place.

**EVERY NEWBIE NEEDS TO KNOW:**

"Bring the brothers home, and sisters home now. Legalize marijuana and take all that money and invest it in teachers and in education. You will see a transformation in America." — *Carlos Santana, musician*

Note that we're talking about nonviolent offenders. If you hire an employee with a history of violence and that worker goes on to hurt someone while on the clock, you could be held responsible!

## PRIMING THE POLITICAL PUMP

If you succeed in turning your enterprise into an ambassador for the movement, your circle of stakeholders may expand beyond into other areas of your state or other states. Business-minded Americans who may have been lukewarm or even opposed to the idea of legalization are willing to change their minds once they see that there is good money to be made in the legitimate industry.

One thing business knows is that you have to have friends in high places to get things done and, if you don't have friends in high places, you better find political candidates to support who are more to your liking. That means hosting fundraisers and showing up at political events. As you invest in the electoral process, those office holders and voters become stakeholders, as well.

Conferences and expos give service and equipment providers an opportunity to connect with potential clients. Credit: VaporDave

## PATIENT & CLIENT RELATIONS

No business can survive without customers no matter how good the idea behind it is. So, if you don't put your clients first, you'll be closing your doors before you know it. The cannabis industry balances on two factors: a culture of cannabis consumers and easy access to the illicit market.

## A UNIQUE & LOYAL CANNABIS MARKET

Cannabis consumers have a unique culture that sets them apart from other consumers. For decades, the only way to obtain cannabis was through the illicit market. When undercover narcs patrolled the streets ready to pounce and no transparency existed to alert consumers of any chemicals that may have been sprayed on the bud (perhaps by the U.S. government), consumers had to trust their dealer implicitly. Once they found a dealer they liked and knew they could trust, consumers rewarded that dealer with unparalleled loyalty.

**EVERY INDUSTRY NEWBIE NEEDS TO KNOW:**

It's a more than a great idea to send some of your hard-earned dollars to trade association and advocacy groups — it's a great investment. These groups are your voice, at a time when very little has been set in stone. For a list of worthy groups, see Recommended Resources.

Still today, an emphasis on trust and loyalty has endured within the community. Do your customers know they can find consistent variety and quality at your facility? If your medicine is labeled as organic, can they trust it to be totally free of pesticides and other chemicals? If you draw them in with specials and promotions, can they trust you to not "bait and switch" them later? Reward their trust with gratification so they can reward you with their loyalty.

Cannabis consumers — especially the "heavy users" who account for a majority of the cannabis products bought and consumed — have a proud history of connoisseurship that strongly affects their purchase and consumption patterns. This is partially a byproduct of prohibition; since dealers are typically sentenced based on the weight of cannabis they were charged with, dealers responded by keeping the customers satisfied with minimal quantity and maximum quality. Growers developed techniques to maximize THC production and boost the terpenes that produce the myriad flavors and scents for which top-shelf cannabis has become famous. As penalties for quantities grew harsher, the price and potency of cannabis rose steadily. For that kind of money, U.S. consumers soon developed a more mature, cultivated palette for cannabis.

***Ancillary*** – adj. ✍

Peripheral yet essential. See Chapter 13.

The California dispensary culture has offered ancillary services to the point that many patients have come to expect offerings of yoga classes, massage and other complementary services from their dispensaries. This "horizontally integrated" model has arisen in response to a system of non-profit ownership agreements and a philosophy of wellness that drives consumer expectations. Medical marijuana patients expect better treatment from their dispensary than from their HMO or local clinic, where long waits lead only to medical care that treats them like a collection of body systems and parts, not a human being. Offer holistic care services to treat the whole patient — and your patients will take care of you, too. These services are tax deductible,

**CANNABIS, HEALTH AND WELLNESS**

True cannabis connoisseurship goes beyond THC. The finest bud doesn't necessarily have the highest lab results; many factors, from the absence of chemicals to the bouquet of terpenes, play a role.

whereas the IRS may not allow you to take deductions for things that are cannabis specific.

## THE ILLICIT MARKET — IN EASY REACH

If consumers get tired of high prices that don't buy high-quality bud to match the cost, many can easily revert to their old dealers. Some regulated cannapreneurs, faced with these facts, may be tempted to try to undercut the illicit market price to compete. This is difficult at best. Legal businesses have no choice but to spend large sums to get their business into compliance — legal costs, taxes, rents, licensing fees, salaries, benefits — and illegal dealers don't. Illicit dealers can pass on these savings to their consumers, while legal businesses trying to match the price cuts will probably only put themselves out of business.

Variety and quality are two keys to success.
Credit: Chris Conrad

**EVERY INDUSTRY NEWBIE NEEDS TO KNOW:**

Remember, most of your patients still have their old dealers on speed-dial. Find ways to compete that the old illicit network can't use.

The smarter strategy is to compete not on price but on quality assurance: gourmet products with brand recognition. Illegal dealers don't usually have access to laboratory testing for potency or contaminants, but the legal market will. By leveraging the cannabis culture of branded connoisseurship, you can provide a product that is demonstrably safe and effective — something your competitors in the illicit market will find difficult to do.

Make sure your budtenders are well trained to take care of client needs, including the special needs of inexperienced newbies.

## EDUCATION IS A MARKETING TOOL

An often overlooked way to care for your customer or client stakeholders is to offer education on how to use cannabis more healthfully. Many cannabis consumers smoke their product by default; however, sometimes it's all they know about or think they can afford. Will your business beat the competition by taking a bit of extra time to listen to your patients, discuss options with them, and introduce them to the myriad ways to consume cannabis safely?

Consumer education also straddles the stakeholder's boundary between service to customer and respect for the broader community, especially when it comes to education on socially responsible cannabis use.

### EVERY NEWBIE NEEDS TO KNOW:

If you're in the presence of children, it's generally a bad idea to be consuming cannabis. If in strictly adult company, it's polite to ask before smoking or vaping. Sometimes the discretion offered by tinctures and edibles can really come in handy. If you're alone, who's gonna stop you?

Do your customers know, for example, when it is socially appropriate to "light up"? Responsible owners ask patrons to refrain from toking in front of neighboring businesses, residences or parks. On-site consumption can solve that problem.

Stoned driving is less dangerous than drunk driving, but with two BIG caveats:

- Naïve users (that means YOU, newbie) can become much more impaired than more experienced users, who generally find ways to calibrate their doses;
- Cannabis and alcohol together produce an impairment that is considerably greater than the sum of their parts. Never drive after taking these two drugs.

What about littering? If other members of the neighborhood find empty bags lying on the sidewalk with your business's name and logo plastered on them, they probably won't know who exactly dropped them there but they *will* know that the litter originated from you. Will you allow the sloppy habits of your customers to pollute your standing within the community, or will you innovate? The Cannabis Buyer's Club of Berkeley (CBCB) in California, for example, keeps a convenient bin where patients can return their used containers to be sterilized and reused with new dispensary product. This has reduced the neighborhood litter around CBCB and also saved the collective some money. The most vital intersection between your customer stakeholders and the broader community, however, has to do with safety on the road. Double parking in the street, sticking out of the parking space, loud music thumping from woofers. Your neighbors will hate that. If one of your customers causes property damage — or worse — it could pose a major setback to your collective's standing in the community.

## EMPLOYEES

**EVERY INDUSTRY NEWBIE NEEDS TO KNOW:**

We use the term "employee" in a broad sense that includes independent contractors.

Equal in importance to the community and patient stakeholders of your business will be your business' workers. There are some in the cannabis industry who work strictly solo and never hire staff, but that means their businesses don't grow beyond a certain size. Cannabis is a labor-intensive business, such that any cannabis business that will grow to any appreciable size will have to, at some point, hire employees.

So like it or not, the cannapreneur of today will likely have to find a way to balance the interests of her customers, her community and her shareholders with those of her employees, who have as much stake in the success of the business as any of these other groups. If this sounds difficult, well, we never promised that success in the cannabis industry would be easy. But for the cannapreneur who can successfully juggle the concerns of each of these stakeholder groups, the rewards can be well worth it.

**EVERY INDUSTRY NEWBIE NEEDS TO KNOW:**

"Employing and promoting women and people of color is the best way to build a successful and sustainable cannabis business." — *Danielle Schumacher, THC Staffing Group*

Employees are the stakeholders who make your business run. Credit: russavia

## UNIONS

The word "unions" acts as a kind of political Rorshach test — most people either love them or hate them. But while we don't want to tell you what to believe about organized labor in general, we strongly advise all cannapreneurs to make and maintain contacts within cannabis worker unions. Even

Badly-named strains create public relations problems for a medical users and providers. Referring to them by initials or substituting an in-house name may be appropriate.

if your company doesn't unionize, maintaining these contacts can be hugely valuable; the union is one of the best sources of well-trained and experienced human talent. Not only that, but if you keep an ongoing conversation with cannabis worker unions, this can be a great way to find out what your employees expect from a worker-friendly company. If you support unions, this conversation can make for a smooth, amicable transition toward a unionized workplace for your company, in which you are seen as a cooperative partner and not an adversary of union interests. If you'd rather not deal with a union, this conversation can be even more important, because it will tell you exactly what your workers want. If you give it to them from day one, what reason would they have to unionize? Treat people fairly.

It's no coincidence that workers who labored in the cannabis industry under prohibition often called their network of fellow laborers "The Union."3 Laboring for decades under constant and mutual threat at the hands of their own government, these workers have developed a sense of camaraderie and *esprit d' corps* that resembles today's organized unions in all but dues and paperwork. It may be only natural for these hardworking employees to more formally unionize now that prohibition is ending; whether you choose to go with that flow or try to find an alternative is up to you. But in keeping with this proud history, it's a good idea to always treat your employees like a union, regardless of whether a formal union is in place.

Now that you've met the stakeholders, it's time to take a look inside yourself in the next chapter as you develop your business prospects.

---

1. See, for example, Proal et al, "A controlled family study of cannabis users with and without psychosis," Schizophrenia Research, January 2014, Vol. 152, Issue 1, pp. 283-88.
2. For a detailed examination of the human toll of the drug war, see Shattered Lives: Portraits from America's Drug War by Mikki Norris, Chris Conrad and Virginia Resner. Creative Expressions, 2000.
3. For a more in-depth exploration of this dynamic, see the award-winning documentary of the same name.

Harborside Health Center stresses professionalism in its design. Credit: HHC

# 10 SURVIVING THE CANNABIS MARKET

Commentators speak of this reform era as the end of a "black market" industry; some go so far as to label the newly legal industry as "white market." Not only is this wording a little awkward in light of the legacy of racism that tinges our nation's cannabis policies, it's also misleading to think of the U.S. cannabis industry in terms of such stark dichotomies. It's evolved; now it's finally getting regulated (and taxed). If one were to talk of the cannabis industry in terms of black and white, it would be more accurate to describe the market as existing in a kind of gray area: while in some ways it's out in the open, in important respects, it still operates in the shadow of prohibition.

Credit: Laurie Avocado

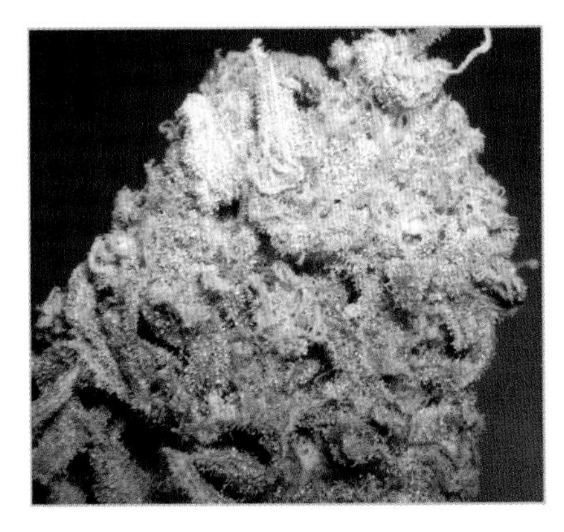

A good trim makes cannabis more valuable.

So how does the newbie cannapreneur navigate through the shadows into the light? First and foremost, anyone who wishes to succeed in the lucrative cannabis market should recognize it for what it is: a community. Secondly you need a solid business plan strategy to guide your operations.

## RETAIL MANAGEMENT

Where will you hang your sign? Location is key.
Credit: O'Dea

No matter whether you're planning a for-profit retail store or a nonprofit collective, your planning and ability to efficiently manage basic business skills can and will make the difference between success and failure for your legal cannabis enterprise. While this may be true of any business, it is especially true in the regulated cannabis industry where it has become increasingly common for state regulators to require license applicants to be able to plan their operations well in advance. Here's a look at what to expect — and what will be expected of you — as you embark into this rapidly emerging field:

## LOCATION IS KEY

The old adage that the three most important factors to business success are location, location and location may be less true now in the age of Internet deliveries and cloud-based services, but when it comes to most canna-businesses, that saying still holds its own. Many local governments have zoning ordinances that severely restrict the commercial and mixed-use zones where people can legally open a dispensary, forcing any

Make sure you're marketing cannabis products only within your home state! They're not legal to ship to other states — yet.

A clean, professional-looking setup at Medicine Man helps create customer interest.

would-be dispensary operator to carefully select locations based on narrow legal zones in addition to cost and customer traffic. If you can't get zoned, try to get a zone variance or waiver. Even delivery services must take care in selecting their base of operations, to make sure they can service a delivery area that contains patients who would be interested in joining their collective and to stay within any local regulations.

Once you've chosen a location and made sure it will pass a city inspection, consider how patrons will find it. You may be tempted to put up the biggest, brightest neon sign you can find to make sure everyone can see your dispensary from the freeway, but it's unlikely that your city, which probably restricts the size and operation of signs within its jurisdiction, will approve such a plan. Check with an attorney or your local Chamber of Commerce to find out what kinds of restrictions exist on store signage, and be realistic about how many customers it will draw in. If your location won't realistically bring in enough customer traffic to cover the costs of doing business, look elsewhere.

## MARKETING

Every business needs a marketing plan. Your potential customers are bombarded with an average of 300 commercial advertisements every day, so your business needs a plan to rise above the noise and make sure that customers who are interested in your product can find you. Here are some of the top concerns: Public relations, advertising, branding, appropriate marketing and patient relations to keep return customers.

## PUBLIC RELATIONS

Some dispensary operators run into a new problem the very moment they select the location for their store: neighborhood opposition. While most of your would-be neighbors will know that the opening of a new medical marijuana dispensary in a neighborhood will neither elevate crime rates in the area nor lead to increased incidence of teen use, there is often a small but very vocal minority opposed to any dispensary in their neighborhood.

Be ready to explain your plan at city council and planning commission hearings. Credit: Sage Ross

One of your earliest jobs, which should be done long before you roll out the new store, is to engage these people head on. Calling a town hall meeting for the neighborhood could work — or it might draw out your opponents and help them find one another. Most of them oppose your operation out of fear, and the most powerful fear is that of the unknown. Pull back the curtain on your plan to dispel such fears. Invite members of your burgeoning collective to talk about their health issues and how medical marijuana helps them.

A more discrete approach is to survey the neighborhood and talk with people one on one, leave them some information about how clean and safe your facility will be, and give them a phone number to call you if they have any questions or problems. That keeps them talking with you directly instead of banding together and going to local officials with unfounded fears and complaints. Come armed with knowledge to

**EVERY NEWBIE NEEDS TO KNOW:**

"If you're entering into the cannabis business, your business is politics. Therefore it's vital to understand that political advocacy has to be built into your business plan." — *Aaron Smith, Co-Founder & Executive Director, National Cannabis Industry Association. TheCannabisIndustry.org*

New research has confirmed that, if anything, the liberalization of cannabis laws leads to a drop in crime. *See, Morris et al (2014) in PlosONE.*

debunk residents' worst fears, such as increased crime or their neighborhood kids getting addicted to drugs.

Help them see that offering a legal alternative for sick patients to access their medicine may actually *reduce* crime. Show them that the competition for patient access may actually help protect kids, since local drug dealers (who may also be offering substances far more dangerous than marijuana) may decide that they no longer want the competition and move on to a new area. Although more research needs to be done, these reasons may be why states that legalize medical marijuana tend to see a slight drop in teen marijuana use after setting up dispensaries.

But while supplying information is a vital goal of public relations campaigns, your primary job will be to listen. Pay close attention; you may learn that there is a relatively straightforward way to address neighborhood concerns. If you find common ground and compromise, you will completely flip the opposition around. Once you include some of their suggestions, they become a stakeholder in your operation, too. Strategic compromise can turn enemies into allies.

**EVERY NEWBIE NEEDS TO KNOW:**

"Get to know the cannabis community by attending meet-ups, grabbing coffee with like-minded people, and building relationships." — *David Hua, Meadow co-founder and Oaksterdam alumnus.*

Finally, remember that the job of neighborhood relations isn't done once your doors open for business. Keep the lines of communication open to ensure a positive relationship with your neighbors.

## ADVERTISING CANNABIS

"Legalize it," vows Peter Tosh, "and I will advertise it." Unfortunately for you and the legendary reggae musician, federal drug law isn't quite so simple. While it is important for any business to advertise their products or services, any attempt to advertise the distribution of cannabis products may provoke the thorny conundrum of federal law. Despite the First Amendment and

major federal reforms that are gaining steam in Congress and the White House to roll back the ban on medical marijuana (see Appendices), it doesn't mean that anyone can advertise it any way they want without repercussions.

Think of the voiceover at the end of an ad for pharma pills or the full page of fine print accompanying prescription heartburn treatments to appreciate the scale and byzantine scope of FDA advertising regulations. Complying with all those rules is expensive and a major reason the U.S. market for pharmaceuticals is dominated by a handful of multi-billion dollar drug companies. Are you really expecting to beat them at their own game?

The issue is different for doctors. The Ninth U.S. Circuit Court of Appeals ruled that despite its illegal status, doctors who "recommend" cannabis are protected by the First Amendment; they just can't "prescribe" it. So while doctors have a right to advertise as do other professionals, care must be taken in how you word your offerings. Furthermore, if a doctor doesn't do due diligence on their cannabis patients, they can be hauled up on charges before the medical board and possibly lose their licenses or face criminal prosecution.

**EVERY INDUSTRY NEWBIE**
**NEEDS TO KNOW:**

A safer, more reliable way to advertise cannabis is through leafly.com or Weedmaps.com, which only make your location visible to consumers in your area.

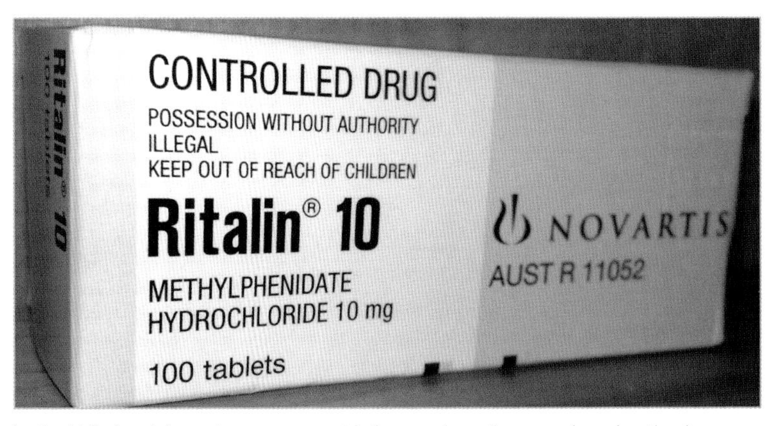


In the U.S., legal drugs have many restrictions on how they may be advertised.

Finding customers for your legal cannabis business can be tricky, but friendly services are here to help. All it takes is a little creativity.

## 420 GIRLS PRESENT A TWO-EDGED SWORD

Amid onerous advertising restrictions, some cannabis businesses have resorted to a time-honored but questionable method of attracting customers: sex sells. At some of the more raucous conventions around the U.S., it may seem like some booths are in close competition to see who can have the least-clothed women hawking wares. There's no denying that such business tactics work in attracting the young-adult male segment of the total market, but consider the cost.

Nationwide, the majority of consumer buying power is in the hands of women, and female cannabis consumers are one of the fastest-growing segments of the legal market. Why alienate half your potential customers with such controversial tactics? It doesn't look medical, it won't draw gay men and many heterosexual men, and women are turned off by the objectification of the female form. So entrepreneurs who think they're bringing in plenty of new business with scantily clad "420 girls" need to consider what — and whom — they might be losing in the process.

Let your cannabis business be a better example for your peers: Hire female employees for their skills, experience and work ethic — not as testosterone bait.

**EVERY INDUSTRY NEWBIE NEEDS TO KNOW:**

"Women are the future of the cannabis industry. They are the fastest-growing segment of the market. Up to this point men have dominated the industry and women consumers have been underserved. Female entrepreneurs understand the needs of female consumers and are uniquely poised to take advantage of this untapped market. The industry, like others, has used sex to sell but it will need to move beyond this for it to reach its full potential." — *Lauren Vazquez, Esq.*

Dr. Sanjay Gupta and his employer, CNN, were instrumental in waking many people up to some of the benefits of medical marijuana, especially for children. Credit: Cubie King

## LOCAL BUSINESS, NATIONAL BRANDS

While there are very good reasons not to advertise the sale of cannabis across state lines, advertising a *cannabis-themed brand* is a different story. The next big thing could turn out to be in franchising, and you might be interested as a franchise licenser or in becoming a licensee.

Harborside Health Center was founded in Oakland, California, but this has not stopped it from promoting itself as a national brand, as anyone who watched the Discovery Channel show *Weed Wars* already knows. Identities are checked at the door to make sure anyone is a legally qualified patient before they can enter and a collective member before they can buy.

**EVERY INDUSTRY NEWBIE NEEDS TO KNOW:**

Remember that most products and services in the ancillary cannabis industry can be advertised in all 50 states. See Chapter 13.

Or consider Realm of Caring, the Colorado cannabis collective specializing in the cultivation and processing of Charlotte's Web, a strain that is high in CBD and low in THC. In a brilliant marketing move, founders Joel, Jesse, Jon, Jordan, Jared and Josh Stanley managed to get their strain featured prominently on a

A more appropriate image for women in the cannabis industry... and a more accurate one.

CNN documentary by Dr. Sanjay Gupta. There are plenty of collectives in multiple states offering high-quality CBD oil for sick kids, but for millions of people who tuned in to CNN, the first name they associate with CBD and epileptic children will be the Stanley brothers. They can't export their product but they can export their brand worldwide.

## Trade Shows, Conferences, Business Associations

One of the many ways that the cannabis community stays connected is through a series of conferences, trade shows and other events held throughout the year. Even in the information age when computers have eliminated paperwork and made our lives easier in every possible way, there is no substitute for meeting your community face to face. For the newbie curious to know where to start, see the list of conferences and trade shows in the Recommended Resources section.

## Beyond Indica & Sativa?

Many believe that sativa cannabis produces a headier, "buzzier" high, while an indica is a more physically relaxing body high.

This view, while popular, has scant evidence to back it up. While the genetic lineage of a plant may determine its physical appearance, the blend of cannabinoids and terpenes that determine therapeutic effects are dependent on many other factors, as well. The available nutrition, temperature, humidity and light all play an important role in determining final plant chemistry.

Consider selecting cannabis based on its cannabinoid and terpene content, which determines its flavor and exerts subtle yet powerful influences on the quality of the high. Here are some of the most significant factors:

- Tetrahydrocannabinol: THC causes powerful subjective effects. The higher the THC levels, the more powerful the high.
- Cannabidiol: CBD mitigates some effects of THC and has therapeutic effects of its own. Medical patients who don't want to get too high should seek out balanced blends of THC and CBD.
- Myrcene: a terpene found abundantly in some cannabis, but also in other plants like mangoes, that can exert a powerful stony feeling in conjunction with cannabis and may cause the stony body high consumers associate with indica strains, according to the Rev. Dr. Kymron deCesare, professor of chemistry at the UC Davis. This is why eating a mango before consuming cannabis is said to intensify the cannabis' effects.
- Limonene and pinene: terpenes found in strains more commonly associated with sativa, and are principally responsible for the buzzy head lift of sativa strains. For a dose of limonene, find a lemon and scrape off the zest of its skin while holding it under your nose. This may wake you right up! For pinene, crush fresh pine needles under your nose for a nice heady buzz, with or without cannabis.

## A TRUSTED HEALER (PATIENT RELATIONS)

Two of the best reasons for a consumer to move his money from his old dealer to a taxed and regulated proprietor is the wider selection and the trusted guidance one can get at a dispensary or other retail store. Train your employees to know how to listen to a client's needs, and stock the kind of product they're interested in, in order to capitalize on this advantage.

Smell is the sense most closely tied to memory; cannabis is the source of the greatest cornucopia of scents and therapeutic aromas of just about any plant species on Earth. Cannapreneurs who learn how to apply these two facts will watch their business blossom. At most successful medical marijuana dispensaries, "budtenders" facilitate a ritual between the patient and the medicine. They give each patient the chance to examine exemplary sample buds up close and invite them to smell the complex bouquets each strain uniquely offers. Any customer who undergoes this ritual knows how powerful it can be!

The wide diversity of cannabis aromas and flavors is far more complex and varied than even the finest wines can offer; one of your jobs as a cannabis entrepreneur is to help your customers discover this amazing variety for themselves. Keep a few jars of coffee beans behind the counter, so customers can take a whiff of the beans to "cleanse their palate" between samples.

The best marketing tool is to keep your clients coming back for more.

## The IRS Nightmare That Is 280E

The federal agency responsible for enforcing alcohol Prohibition was the U.S. Treasury. The feds arrested Al Capone on tax evasion charges. So maybe it shouldn't be a surprise that, in recent federal raids, Internal Revenue Code agents seem to be just as ubiquitous as those representing the DEA or FBI.

Now taxes are once again being waged as a law enforcement weapon: against medical marijuana dispensaries. Only this time, the tax man has a new, potent tool what wasn't around in the Roaring Twenties — IRS code section 280E.

Using 280E, the agency has embarked on a nationwide campaign to tax legitimate medical marijuana businesses out of existence. This obscure section of the federal tax code creates a deliberate double standard. Practically any business in America can deduct ordinary expenses like rent, utilities and human resource costs from their taxable income, but 280E explicitly denies such benefits to "traffickers in controlled substances." Originally drafted in the 1980s to battle cocaine cartels, the Treasury has been wielding it as a club against state-legal cannabis businesses when other charges won't stick.

Don't let your cannabusiness be the next IRS target. A few preventive measures can be the best solution to a taxing problem. Make sure you consult with an attorney or CPA who specializes in 280E before setting up a state-legal business.

## Return On Investment (ROI) Analysis

It might be easier to remember the return on investment formula as return *over* investment, because that's the mathematical formula you apply:

*ROI = return (i.e., revenue) / investment*

ROI lets you know how productively you're spending your resources. For example, suppose you have an indoor grow and an outdoor grow. Indoor bud consistently sells for $2,500 per pound; outdoor goes for $1,500. You might be tempted to put all your work into the indoor grow next year; but make sure to calculate the ROI for each option to see which is the better investment. Indoor is generally more expensive to produce, so in this hypothetical:

*investment / lb. (outdoor) = $50 amendments + $20 water + $10 electric + $120 labor (cultivation) + $200 labor (processing) = $400*

*investment / lb. (indoor) = $100 amendments + $40 water + $300 electric + $160 labor (cultivation) + $200 labor (processing) = $800*

Your outdoor grow will probably need much less in the way of amendments if you use sustainable farming techniques and care for the land. Similarly, you can cut the water use in half if you mulch over your plants' soil or try using less water-intensive hydroponic techniques inside. Labor is less expensive during the season, since the outdoor grow doesn't require as much careful attention to getting all the environmental variables right, but this advantage disappears during harvest season, as the cost of trimming and curing cannabis is about the same regardless of whether the product was grown indoors or out. The biggest difference is your electric bill: outdoor bud will only need motion detectors, the occasional oscillating fan and lights for your trimming crews at harvest. Indoor plants soak up electric light as if their lives depended on it — which they do. Plug the per-pound cost into the formula, and:

*ROI / lb. (outdoor) = $1,200 return / $400 investment = 300%*

*ROI / lb. (indoor) = $2,000 return / $800 investment = 250%*

Gather your own data and track it by strain and technique to get the tools to maximize your garden potential — by focusing on highest ROI.

# 11 BUSINESS OPERATIONS

What goes on in a dispensary's lobby is challenging enough; what goes on in the back room can be crazy. Newbie or not, when it comes to the logistics of operating a dispensary, there is precious little margin for error, and consequences can be dire. If you are serious about going into the legitimate business, you need to do all the things any other legitimate business does, such as getting permits, licenses and inspections, and paying quarterly taxes, etc. We don't have space for all that here but, due to the added complexities of cannabis industry compliance, we strongly advise you to consult with a lawyer and accountant to see how is best to proceed: incorporation, non-profit, sole proprietorship or something different. Likewise, there is a lot of paperwork and registration at the state and local levels. Take care of the paper process and code enforcement in a timely manner to avoid fines and delays.

Let's look at some of the things that make the cannabis industry different.

## INVENTORY CONTROL

Losing track of inventory is bad for any business, but in the cannabis industry, it can mean an orange jump suit. Since 2012, legalizing states have followed a trend of "seed-to-sale" tracking for all cannabis products in a collective or other business'

> **EVERY INDUSTRY NEWBIE NEEDS TO KNOW:**
>
> Inventory tracking can be much easier with the help of automated software designed for the cannabis industry. Two leading programs are MJ Freeway and BioTrackTHC.

inventory, and the federal Department of Justice has made it clear that their tolerance of state programs only extends to operations that strictly monitor inventory to prevent diversion to the illicit market. Thus, taking extra care to keep track of your cannabis inventory isn't just good business, it's the law.

So break out your serial numbers and bar codes. Here's what you need to do to make sure your cannabusiness is both compliant and financially sustainable.

## THINK SECURITY

Cannabis is so valuable that there is a risk of theft. Be sure to have adequate security: solid walls, maybe a guard, surveillance cameras, motion detectors and lighting. Make sure outer walls are solid, not hollow. Keep things out of sight. Think about getting a safe for security.

## PURCHASING & PROCUREMENT

The first step in successfully managing inventory is to always make sure that you have the products your patients want, when they want them. This is no easy task, especially for entrepreneurs who are new to the rapidly changing industry. There are two principal ways that legal cannapreneurs solve this.

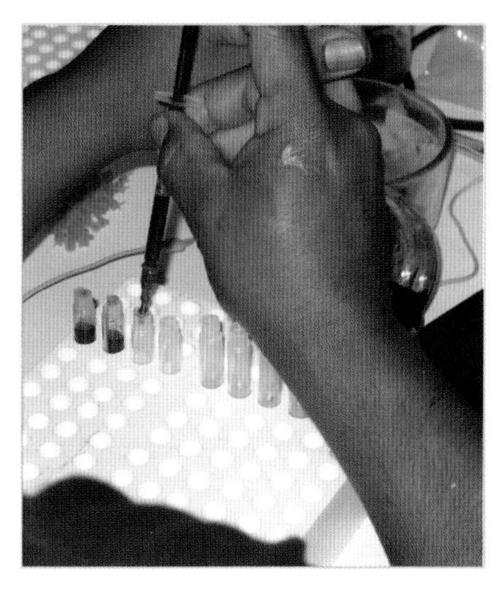

Increasingly, cannabis consumers and regulators are demanding lab-tested products in measured amounts.

Some retail outlets find it easiest simply to grow and process their own cannabis — some states mandate this approach. Colorado, for example, requires 70 percent vertical integration for retail store licensees, meaning that anyone who has a store under the state's adult-use legalization program also needs a plan to grow and process their own cannabis. Even in states that don't mandate this kind of model, many retail operators voluntarily choose to grow some or even most of their own medicine, because of several advantages this method conveys:

- The retailer has tight control of the product right from the beginning of the plant's life, allowing for more fine-tuning of the final product than she is likely to find any other way.

- The retailer has an easier time proving lawful chain of custody if the plant and product have not left her control.
- The retailer can obtain the cannabis at a lower cost because she has salaried staff and doesn't have to pay the markup of the growers or any middlemen, such as cannabis brokers. She can pocket the difference as profit, pass the savings on to her customers, or both.

Of course, vertical integration has its downsides, too. The retailer will need at least one person on staff who knows how to grow great cannabis and is willing to make the task his exclusive full-time job (see Chapter 12). She has to internalize the risks faced by all cannabis cultivators. She has to choose which seeds or clones to grow by predicting the fickle tastes of the consumer market three to eight months later, when the crops will be ready to process. Will she lose the crop to spider mites, to a catastrophic power outage, or to theft?

Farmers have always been the victims of systemic risk, and cannabis farmers are no exception.

Weights fluctuate a few percent, depending on ambient humidity.

Many retail outlets understandably decide to skip the extra work of growing their own product, at least at first (you may want to revisit this as your operation grows). Others may prefer to grow most of their own supply but supplement the in-house supply with a greater diversity of products, like medicated edibles, that aren't practicable to produce in-house. Either way, every retail outlet needs a trained purchasing department to source clean, high-quality products through trusted vendors, and issue sales contracts and licensing agreements.

At this writing, it's a buyer's market for unbundled cannabis, so don't be surprised if growers start to contact you *en masse* as soon as you hang out a shingle or put a pin down on Weedmaps.com. Your job will be to decide how to respond. Every grower will tell you that they are the best on Earth. To succeed at a retail business, you will need some way to separate the weed from the chaff.

## ALLOCATION

Most of the time, when retailers buy cannabis through their purchasing department/vendors or even when they grow their own, they receive it in the form of cured, unbundled flower. Typically, growers offer their produce to dispensaries by the pound, with each pound wrapped individually in an airtight oven bag. Once the purchase is made, it becomes the retailer's responsibility to process the cannabis into consumer-friendly sizes and forms.

- Unbundled flowers remain the most popular cannabis form for consumers to purchase, although that may soon change. To meet this demand, retailers break their wholesale pounds down into ounce, half-ounce, quarter-ounce, eighth-ounce (the ever-popular "eighth") and gram sizes. Processing is minimal, although the retailer will typically remove excess leaf and stem material, and break down larger colas to smaller buds for their eighth-ounce and gram-size offerings.
- Pre-rolls/joints are popular with retailers and consumers alike. Consumers appreciate the convenience of having a cigarette already rolled to smoke, and retailers appreciate the opportunity to add value to their broken or unbundled flower. Pre-rolls are a great way to finish out the bottom of a wholesale pound, which may only consist of small, unsellable buds after all the prize nuggets have been gleaned; these "house cones" allow dispensaries

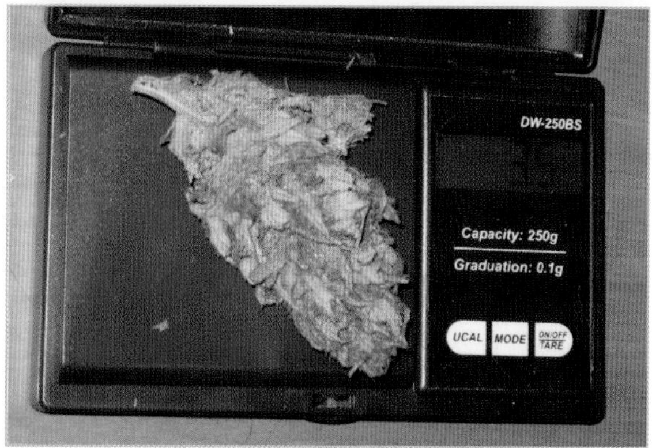


Eighth-ounces, one of the most common quantities sold in the industry, can vary greatly in appearance based on bud density.

an easy way to sell buds that are fine in terms of potency and flavor but lacking in aesthetic appeal, sometimes mixed with trim or kief.

- Concentrates (hash, wax, honey oil) are quickly growing in popularity among consumers, but few retailers make them in-house. Processors and retailers can extract value from the piles of leaf, shake and trim that soon start piling up in a grower's inventories. Some techniques, such as supercritical CO2 extraction, can be expensive. Others, like butane BHO extraction, are dangerous processes best outsourced to specialists. Dry-sifting kief, water-extracting hashish, dry-ice kief or wax- and ethyl alcohol-extracted oils can all be undertaken safely and inexpensively in-house.

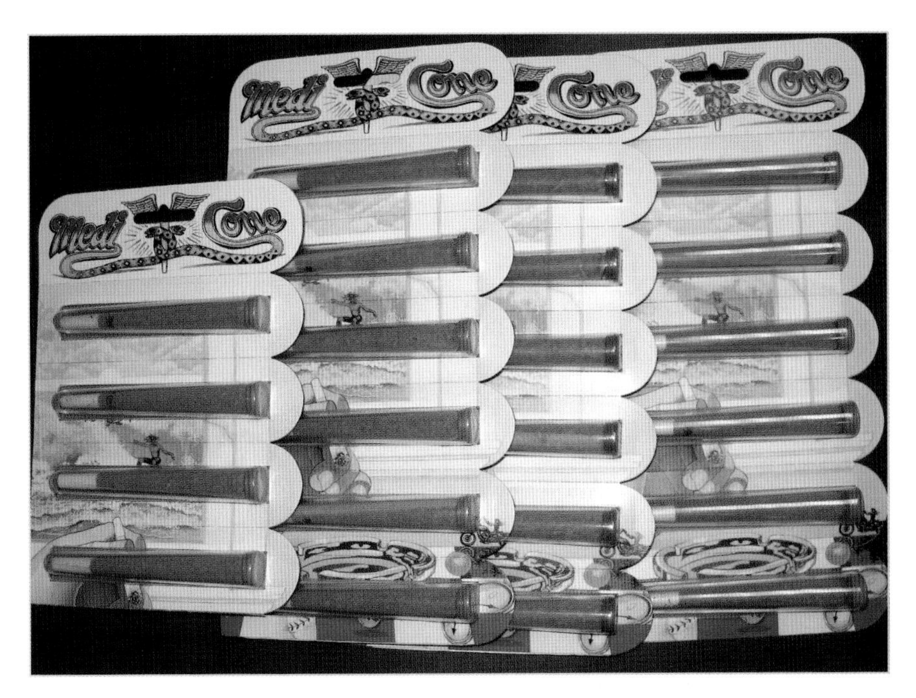


Leftover bud, trim and kief can be blended into pre-rolled joints, which many consumers appreciate for their convenience.

## FRESHNESS & STORAGE

Although cannabis has some incredible aesthetic and therapeutic virtues, they don't last forever. Sadly, cannabis is perishable. If cured and stored properly, it can last for years. But daily handling by retailers will accelerate the degradation process, so take care that every offering on the menu is as fresh as possible on the date a customer purchases it. Ideally, you should never have any inventory more than two to three months old. Edibles have a shorter shelf life and you need to keep your supply current.

One of the most essential ways of solving this problem is to make sure that cannabis products are always properly stored. Stock newer products in the back of the display and

**EVERY INDUSTRY NEWBIE NEEDS TO KNOW:**

Cannabis stores best in a cool, dry environment, totally dark and in an airtight container. Edibles and beverages need to be refrigerated and carry a "use by" date. As this date approaches, consider using the product as a promotional item for loyal customers, an incentive to new customers, or a compassion product for clientele who face financial hardships.

For security as well as freshness, many retailers keep their cannabis products in a cool, dark safe or vault.

dispense older items first so they don't get stale. Equally important is the necessity of tracking inventory by age and retiring out inventory that has passed its expiration date. Make sure to mark the date of each wholesale purchase and assign the inventory a batch number that you associate with every ounce, eighth, gram, etc., you process from the batch. This will allow you to track the age of each bag so you know when it's time to run a clearance sale, break down into pre-rolls, or pursue another liquidation strategy. Refrigerate your edibles.

Wax, oil and "shatter" are among the purest solvent concentrates available, but the process can be hazardous and they should only be prepared in a licensed extraction lab.

Track your inventory's aging by SKU or other identifying number, so you know which items stay on the shelves longest and which get picked up quickly. This is a great indication of what kinds of products your customers like best! It will also help in planning your cash flow, because you'll know how often you need to make future wholesale purchases with greater precision.

## DELIVERY BUSINESS

If all the discussion of the myriad costs of running a dispensary or other "brick and mortar" retail outlet has left you with a severe case of sticker shock, consider a delivery business. Rents will be much lower, as you won't have to pay for a prime retail location. Typically, human resource costs will be lower as well, because of the reduced staffing needs of a mobile delivery service. Even if you have a dispensary, adding a delivery service will allow you to offer the convenience of home delivery to your patients. But before you rush out to buy a couple pounds of cannabis and a pin on Weedmaps, there are some important facets of running a legal delivery business to consider. What is the status of the vehicle and its insurance? What specific documentation does your driver need to carry to show that they are legitimate? How do you protect them?

Cannabis is compact and light for easy delivery. Credit: Lbking

## INVENTORY CHALLENGES

Sourcing cannabis is a real challenge. If you produce your own you have all the risks of the grower in addition to that of a retailer. If you don't grow your own, you may run out or have to rely on the black market at some point, which could jeopardize your license.

While in some ways running a delivery service is simpler than running a dispensary, in one very important regard it can be much more complex: handling inventory. A dispensary can keep all its medicine in a single, secure location; a delivery service doesn't have that luxury. While it is a good idea to keep all of your collective's medicine in one secure location at night, during business hours at least some of that inventory has to be entrusted to your drivers in the field. Some delivery services keep a tight leash on driver

## What Is Your Delivery Area?

Naturally, you may want to cast your delivery area as far and wide as possible on day one, under the reasoning that you want to attract as many customers as possible. That's a fine goal, but consider the downside: suppose a patient takes you up on your offer, requiring you or one of your drivers to meet her 50 miles from your base of operations to make a $40 purchase. After you pay for the gas, auto maintenance, tolls and hours of time your driver spent in traffic to get out that far (and back), it's probable that these costs will have eaten away all of your profit — and possibly give you a loss on the transaction. Even worse, this small purchase might end up tying up your only available delivery driver, forcing you to turn down much bigger transactions that may be much closer to home.

That is why all smart delivery service operators define their delivery area by the area they can actually afford. To figure what your profitable delivery area would be, tally all the costs of making a delivery, and prorate them by the mile. Note, too, how much time it takes, making multiple trips during different times of day — a trip during rush hour will almost certainly take much longer than a trip at 10 in the morning.

inventory, and only carry exactly what the patients have requested in advance. This has the advantage of keeping the tightest possible control of your inventory but the disadvantage of losing out on order flexibility and some great upsell opportunities.

For these reasons, many delivery services supply each driver with a small amount of every major item on the menu at the beginning of the shift. This way, delivery collectives can strategically place drivers in different parts of their delivery area, each with a small supply of whatever a patient may order. If one driver runs out, the closest driver can resupply them quickly.

These are all great upsides, but this model comes with a significant downside: the risk of theft or other loss. There's a reason why most pizza delivery companies don't allow drivers to carry more than $100 in company cash. But a single cannabis order can total several times that much. Cannabis delivery drivers are

There's a fine line between a consumer who likes to order in bulk and one looking to deal to his friends on the side. Use your judgment.

In Los Angeles, a person can get their doctor's approval on their way to the beach, leading to more than a million qualified patients and a robust network of dispensaries. Credit: Adam Jones.

usually much more discreet than cars with their well-lit pizza placards emblazoned on the top, but the risk of theft still exists. Carrying a significant quantity of high-grade cannabis products runs the risk of losing product and, much worse, it could put your driver's safety in jeopardy.

Theft is not the only way that carrying inventory around in cars can lead to loss; many cannabis medicines are volatile and can degrade quickly in a car trunk on a hot summer day. A successful cannabis delivery service must find a workable solution to these challenges.

"The cannabis entrepreneur must understand that, at this point, the entire industry rests on a legal house of cards. It is primarily the political strength of the reform movement that keeps the Justice Department from shutting down the cannabis industry in a short series of lightning actions. For every dollar invested in the business, at least a dime should be set aside to support the reform movement's political activities." — *Eric Sterling, Esq.*

## Bulk & Wholesale Stock

Suppose you want to make a large purchase, perhaps to take advantage of a generous discount offer, but you know that you won't be able to sell it all in three months at your current volume. Is the offer really as good as it seems if your customers don't like the product as much as you think and you get stuck with a bunch of rapidly-aging product that you can't move? Even if your customers do like the product enough to pay a good price, is it worth tying up limited cash reserves in inventory that may not be easily liquifiable back into cash? There may be a lost opportunity cost later, when an even better deal comes around and you don't have enough cash on hand to take advantage of it.

But if you do decide to buy in bulk, make sure you process only as much as you need on hand — say, a week's supply — and keep the rest as undisturbed as possible in long-term storage. This will help keep your inventory as fresh as possible.

## Paying Out

In the cannabis industry of today, payments — whether to employees, landlords, or partners — can often be on flexible terms. Some of your colleagues in the cannabis industry only accept cash, but others are willing to accept payment in kind, crop shares or work trades. Get creative in your negotiations; finding a deal everyone can live with may be easier than you think.

- Short on cash at the end of your season? Ask around for a work trade to help get your harvest in. Perhaps your neighbor's late-blooming indica garden won't be ready for harvest until two weeks after yours; see if you can take advantage of the time gap by getting her to help with your busy harvest now in exchange for a promise to help her out two weeks later. If you keep an eye out, you might spot a win-win situation.
- The practice of consignment has grown in popularity at the purchasing departments of collectives, especially when they're working with a new strain or new grower. Today's retail shops often take unproven inventory on a consignment basis to reduce their risk in the event it doesn't sell well; growers get paid as their product moves.
- Have you found the perfect spot for your garden but can't afford the rent? Ask the landlord whether they'd be interested in a crop share, in which a portion of the harvest (typically a third) is given over to the landlord when it comes in. This reduces your up-front costs, as well as your risk — if you lose your crop to mites, at least you won't owe rent.
- Many workers are willing to accept payment in kind, especially if you grow quality bud! Some farmers offer their trimming crews an ounce of finished bud for every pound they trim, instead of the standard $200 per pound cash. Since cannabis trimmers are often cannabis consumers as well, you may find a fair number of competent employees willing to work on these terms.

## LOCAL REGULATIONS CAN CAUSE BIG HEADACHES

As significant as these challenges may be, they may end up paling in comparison to the challenge of satisfying local regulations. While patients tend to love the convenience and discretion of deliveries, many knee-jerk politicians at the local level dislike them because they are more difficult to regulate. A dispensary, with its stable street address, can be easily monitored and made amenable to licensing and inspection requirements; but, from a city council's point of view, taming a long list of delivery services operating within the jurisdiction can be like trying to slay a many-headed hydra. Thus, cities, counties and even states have begun to pass bans and restrictions for delivery services while allowing brick-and-mortar dispensaries to operate. In Nevada, for example, you can't operate a delivery service unless you hold a dispensary license.

Local governments pass a wide range of convoluted ordinances regulating deliveries, with terms and rules all over the map. Some local rules even violate their relevant state constitutions, so, if you are considering starting a cannabis delivery business, it is vital that you consult with an attorney before beginning operations. If you find a competent lawyer who specializes in the cannabis industry, the advice you receive will be well worth the cost.

In sum, do your best to follow local regulations and remember, first and foremost, always put the needs of your patients and clients first.

## Other Considerations

While the previous considerations are the most important to any cannabis delivery business, there are a number of other factors that must also be considered:

Driver pre-screening requirements: Will you require your drivers to pass a criminal history background check before hiring them? Some jurisdictions require this. How much insurance coverage will you need them to carry on their vehicle? At what point do you disqualify the use of a prospective driver's vehicle based on age, state of repair or cleanliness? Many patients place a premium on the kinds of presentation details that signal a high-class service. How will you make sure your drivers live up to the high standards patients demand?

Discretion: Many patients will choose a delivery service over a dispensary because of the discreet service deliveries can provide; but if you fail to anticipate the kind of discretion that patients demand, you will lose this advantage. Meet the expectations of your patients with low-key, professional service and you will be rewarded. If you show up looking like a drug dealer, however, you shouldn't expect to get any repeat business.

Special patient needs: Some patients call a delivery service simply because they are physically unable to get to a dispensary. For the very sickest patients, extra care must be taken. If the patient or her caregiver gives specific instructions on what to do before and during a delivery, make sure you follow them to the letter.

# 12 GROWING CANNABIS FOR FUN, MEDICINE OR (NON-)PROFIT

There are many different ways to grow cannabis, and people have various motives for growing it: some grow it for their own use and to save money; some do it because it is the best medicine for their health issues; some plant it because it is a tried and true way to make money. Some grow for their friends or a small collective; some grow in massive industrial warehouses. Some grow outdoors, some indoors, some in greenhouses.

All of them start by choosing a location — any place from a back yard or spare bedroom to outdoor fields, commercial warehouses and acres-large greenhouses. Newbies will need a discreet place with constant access available.

Credit: Chris Conrad

## LARGER SCALE CONSIDERATIONS

Perhaps you have some notion in your head that the best reason to leave the city life and become a pot farmer is to avoid all the tedious spreadsheets and market analyses that come with modern business plans.

Perhaps you planned to go into cannabis farming so you wouldn't have to have a plan.

Sorry. A hundred years ago, perhaps, the majority of American farmers operated without a business plan, preferring instead to follow the rhythms of the land. It's a life cannabis growers only dream about. In waking life, the illicit market is a road to riches or prison, as the legal cannabis market is a mire of regulations. In fact, growing often combines the headaches of dispensary management with the added vagaries of agriculture. Some states require that you operate as a non-profit corporation. Growers must anticipate future markets — which strain of bud will be hot five months from now, when your harvest is ready?

Add to that the variable of security. You have to protect your investment. To avoid neighbor problems and odor complaints, most grows need to be either very remote or well fenced off, or both, with motion detectors, video surveillance, etc. Some localities require a security guard on patrol 24/7.

So what's your plan? What market will you target? How will you keep costs down? How will you mitigate risk? What are your strengths and limitations? Are you growing outdoors, indoors or in a greenhouse? How big will this garden be? Which techniques will you use? The degree of complexity to your grow and the number of hours you put into it will largely depend on what kind of grow site you intend to set up.

- An outdoor grow can be less expensive and less work but, ironically, it requires much more planning in the manner of traditional agriculture. A small number of plants can yield a large amount of beautiful, potent buds, some of which can compete with the best of indoor weed. In fact, many cannabis consumers categorically prefer the clean taste of organic, sun-grown cannabis to the often artificial or chemical taste of indoor hydroponic. Many farmers find that investing in the land through organic and biodynamic techniques can reap dividends and make life easier by enriching and maintaining the soil throughout the season.
- A greenhouse grow can be a great choice for growers who wish to combine the precision control of an indoor grow room with the bud-bursting power of the sun, but be advised that a successful greenhouse can often be as much work as an indoor grow, particularly if you augment it with lamps or force flowering.

This automated greenhouse uses supplementary lighting and light deprivation tarps to precisely control cannabis flowering all year around. Credit: Chris Conrad

Even so, some pioneering companies offer partly automated greenhouses to ease the workload.

- An indoor grow room, let us assure you, can be just as complicated and time-intensive as running a retail shop, especially if it grows to any appreciable size. By taking their plants out of their natural environment and placing them in an entirely artificially constructed environment, indoor growers have to assume responsibility for each and every one of their plant's needs — and for making sure they are safe and nourished 24/7. Many indoor growers, running several cycles per year, see the amber glare of their grow lights far more often than the brilliant light of day. For all this, indoor cultivation comes with a tempting upside: the highest wholesale prices in the industry, for those few gifted growers who get every meticulous detail just right.

No matter which option you choose, you will benefit handsomely from buying grow books that specialize in the technique you want to pursue and putting together a cannabis cultivation business plan. Address potential problems in the planning stages. Are the ceiling height and ventilation adequate? Does the site need a power upgrade? Where does spilled water run to?

## START SLOW & KEEP ON GROWING

Once they understand the kind of profit that can be made from growing one pound of high-quality cannabis, many growers understandably want to aim to grow a thousand. Resist this temptation. It could prove to be your undoing.

The finest cannabis is generally produced in small batches, so the grower can give each plant the artisanal care and attention they need to attain their maximum potential. The operation may grow organically from there, but for cannabis farming, you are looking for sustainability. To find that magic point, any cannabis farmer must know his or her limits.

So let's get you growing, newbie. (For more advanced instruction, seek out grow guides by Jorge Cervantes, Ed Rosenthal and Kyle Kushman. See "Recommended Resources" for more information).

## GROWING FOR BUD

If you don't know how to determine plant sex or assure that your garden is feminized, now's a good time to review the first chapter as to how to identify and separate staminate male plants from pistillate females. Female cannabis grown for its bud and resin is best kept *sinsemilla*, or seedless.

The yield from any given female plant is determined by its variety and its size. Before the plants start to bloom, they go through a vegetative period during which their sex is not readily evident. The larger the plant grows in its vegetative period, the more branch nodes there are to form flowers once it starts showing its sex. However, the larger the plant grows, the bigger a shadow it casts and the more difficult it can be to work on. In the right

Credit: Chris Conrad

Credit: Chris Conrad

Once this plant is topped, these side branches will take off in growth.

This plant has been pruned at the top to encourage more branching.

Once topped, cannabis branches into bushier growth.

conditions, a single plant can grow more than a dozen feet tall. Do you really want to be standing on a ladder for hours on end looking for bugs, mold or male flowers? If the answer is no, then you probably want to grow smaller plants that yield less, but harvest more of them, more often.

To control the plant's size, prune attentively throughout its vegetative growth period. First, top the plant at an early stage, to trigger branching growth rather than height. More branches means more flowering tops, and that is where the sticky resin of terpenes and cannabinoids will be excreted through the resin glands to form into trichomes. Clones are made from branches, but the same theory applies: taking off the central branch and a few branch tops or bending the branches will redirect the energy flow of the plant, control its height and result in more buds, although somewhat smaller.

After growing a certain size, plants track the hours of light they get in a day to trigger flowering mode. Outdoors, the switch occurs after the summer solstice; indoors or in greenhouse gardens, it can be induced at any time. Now's your last chance to cull any stray males! Once flowering starts, the blossoms cluster together into buds. As the buds grow, they produce terpenes — the aromatic molecules that give plants their varied and respective fragrances. Terpenes can be very beneficial, and the release of cannabis fragrance into the world is generally a very positive thing, but since it is subjectively detectable, it gives neighbors something to complain about and could draw unwanted attention from authorities.

Although there are few plants, multiple strains are growing. Excellent results can be attained by using bags of amended soil, as pictured here. Moisture escapes through the sides of these cloth bags, and they should be covered or surrounded with straw, when possible, to retain water.

Outdoor gardens can have plenty of area to fill with their canopy.

As the plants near maturity, pull away some fan leaves to allow more light to reach in to increase the lower branches' yield.

## OUTDOOR GARDENING

Growing cannabis outdoors is the natural way and it can be done in any number of techniques. Neighbors grow a few plants in their back yards. "Guerrilla" growers hide cannabis among trees. You can till and fertilize the soil like you would for other farm crops, you can grow it organically, you can grow it in mounds, you can grow it in soil bags or planter boxes, you can dig pits and trenches and fill them with amended soils, and you can build trellis frameworks around them to support the branches and allow denser growth on the plants; these may consist of bamboo, wood, nets and wire mesh, usually on a 4-inch by 4-inch grid.

The different growing styles, garden health and specific genetic phenotypes all help determine the crop's ultimate yield. Most outdoor growing is done in soil; either planted into the earth in amended soil with drainage, in wooden beds, in plastic or ceramic pots, or in grow bags holding anywhere from less than a gallon to hundreds of pounds of soil. Regardless of size, the blends focus on three principal nutrients: N (Nitrogen), P (Phosphorus) and K (Potassium). Every grower seems to have their own favorite formula, and different strains seem to have their own preferences as well.

If you are growing a few plants or a medium-size patch of cannabis, using a watering can or a hose with a variable nozzle or "wand" to water the garden gives the gardener a chance to spend more time with the plants and

These "screen of green" plants are grown with a grid used to spread the branches and support their weight as they become filled with swelling flowers.

Patients pay close personal care and attention to the health and quality of their bud.

observe their health. The plants need good drainage, and in a bigger garden, the use of drip irrigation to water the plants becomes more important. It delivers water right to where it is needed, watering the roots deeply and saving water consumption by reducing surface area evaporation. Cannabis plants love nutrients and water, but if you give them too much of either or both, you might end up with root burn or root rot that strike quickly and relentlessly just as your plants are looking great. Don't ignore the microbial soil structure. Using compost tea/worm and mulch is important to build the subsoil and grow the mycelium networks that keep the soil rich and plants healthy. Mulch also helps control root pests and makes for more efficient water use.

The great thing about growing outdoors is that you get to use free sunlight to grow the crop, the natural breezes fan your plants, and you have access to the earth's support networks to sustain the crop. This is especially helpful if you are working toward organic produce and sustainable agricultural practices. In general, outdoor crops are limited to a single harvest per year, but by putting a makeshift greenhouse over the canopy and using a technique known as "light deprivation," you can get an extra harvest.

## GREENHOUSE CULTURE

Greenhouse growing is essentially similar to outdoor growing; the two systems frequently complement one another. People start their plants early in a greenhouse, with or without lamps to extend the natural hours of illumination, and sometimes move

During its vegetative growth cycle, cannabis prefers a NPK mixture of 10-10-5, and for flowering 5-25-9.

Few growers can afford a greenhouse this elaborate, but glass is more secure and insulative than plastic. Credit: Wiki: Alvegaspar

them outdoors. Gardeners who use light deprivation for a mid-year crop often use a hoop greenhouse to make it easier to slide the tarp over the plants, blocking the light without battering upper branches. Growers may move a few plants into greenhouses in the fall to finish up if the buds haven't fully ripened when the weather turns. Greenhouses are also useful for both sinsemilla production and breeding purposes because they help ensure that stray pollen from unknown male plants doesn't get into your crop.

Greenhouses can be of plastic or glass, but the hoop style has become more popular in recent years. If using glass, one can be more discreet by using textured or tinted glass on the lower panes to obscure the appearance of cannabis. For plastic greenhouses, there are various thicknesses of sheeting that are available, and many kits ready for you to buy and install to get going fast. The contained environment helps protect the plants from animal pests, temperature extremes and music festival attendees.

Like outdoor gardens, greenhouse gardens get free sunlight, but several differences show up right away. The greenhouse conserves moisture but also traps heat, so it is a good environment to use as a nursery but it can also overheat the plants. The height and shape of the greenhouse restrict the

The Grobots system places separate greenhouses inside a larger greenhouse to flower one section of the garden at a time. Light-blocking tarps are guided over lines to easily cover the plants and force flowering.

spatial development of the plants. If you keep the greenhouse enclosed, you can go ahead and discreetly filter your exhaust.

Greenhouse growers have their choice of where to root their plants. Growing in the earth has been tried and true through the millennia. Soil in pots and bags can be put into standard sizes of trays that capture water for the plants and prevent leakage onto the floor, which could result in mold and other problems. Pots or beds can be put on wheeled tables, enabling easy access and allowing storage space beneath the plants.

Like indoor gardening, a greenhouse better lends itself to non-soil growing media like hydroponic or aeroponic gardens. Rather than the roots taking up food from the soil, they get nutrients fed to them dissolved in water as they grip textured materials such as rockwool or coco bark. In aeroponics and deep-water culture, the roots grow through a basket without any medium.

## INDOOR GROWING

If you thought you were using a lot of electricity with that enhanced greenhouse described above, wait until you move all the way indoors. Of course, windows, skylights, sun tubes and reflective surfaces can be utilized to tap into the sun's light during daytime hours, but basically the indoor grower is trying to replicate the great outdoors in an enclosed space. Just replacing the sunlight alone is costly. The yield per lamp has doubled twice since the 1990s, but even at a four-fold efficiency rate, growing quality cannabis without the sun casts a large carbon footprint at considerable expense.

You want your lamps to hang close enough to the plants to get the most lumens for the energy, but not so close as to burn them. The farther away the bulb is, the more its light dims — and the more your productivity drops.

Sometimes one lamp is all you need to get the job done. Credit: Chris Conrad

Indoor grow rooms can vary in size from a single lamp to a large warehouse, like this Colorado garden. Credit: Medicine Man

Adjust the distance by hanging the lamps on chains to raise and lower at will as the plants grow in height. The tables can be elevated and lowered as well, sometimes using telescoping legs or simply by setting soil planters onto boxes. To capitalize on the light your lamps are generating, you want the surrounding surfaces to be reflective, too. Sometimes that means using mylar sheets but oftentimes simply painting surfaces white will work just about as well. Likewise, aiming lamps to shed their light across the same plants from different directions improves photosynthesis.

Because of the intense lamp heat in the room, there must be good air circulation. Carefully prune the lower branches of the plants to allow for better flow of air and light, which can only penetrate a foot or so through dense garden canopy. Plant tops can be pruned judiciously to keep the canopy surface level, or alternatively the growing containers of shorter plants can be elevated from below. Using a plastic grid to spread out and support the branches of the plant will help maximize the efficiency of the light distribution. Another way to increase canopy surface exposure to lamps is to stack the gardens vertically and likewise stack the bulbs into columns.

Each 1000-watt HID lit 12 hours per day for flowering will typically produce a pound, although many factors can push that average up or down. Credit: Chris Conrad

Even a carefully controlled environment needs human oversight to make sure all systems are working and no pests have invaded the garden.

## THE FLOW OF THE GARDEN

There are two lamp cycles used for indoor gardens. A cycle of 18 to 24 hours turned on per day prevents female plants from blooming and allows them to grow for years. Using this illumination pattern, clone growers maintain mother plants, propagate starter plants and grow vegetative plants to whatever size is desired before putting them into bloom.

For the flowering cycle, growers switch to 12 hours on and 12 hours off. Getting extra light in the vegetative phase is not a big problem for the plants, but interrupting the 12 hours of darkness needed by blooming plants can throw off the crop and play havoc with its yield.

Controlling temperature and moving air through the indoor garden are very important. Use carbon filters to remove odor before the air is discharged outside. In order for that to work, you need a negative airflow. That means that if you open a door or window, you feel air being sucked into the room from the outside, not blowing out through the opening. This funnels all the air from inside through the in-line filters and prevents odor from escaping. Keep room temperatures in the range of 70 to 85 degrees Fahrenheit. It is natural for the room to heat up when the lamps are on and cool down when they are off, but avoid extremes. The range for humidity varies over the life of the plant. For seedlings and to help root cuttings into clones, 80 to 90 percent humidity helps keep the plants from drying out. During the vegetative growth phase, bring that down a notch to the 60 to 70 percent range. When in bloom, you won't want to risk mold, so step it down again to 40 to 60 percent humidity.

Whether you grow in soil or hydro, keep the pH of the medium and water slightly acidic — around 6.5 pH — and check up on it from time to time. The water coming out of your tap is already acidic or alkaline to start with, so start by checking that level before making any adjustments. Since rockwool is alkaline to begin with, you need to bump down the acidity of your water to about 6.1 to counteract that. For soil, use soil sulfur to make the soil more acidic or lime to make it more alkaline. For water solutions, there are many pH-up and pH-down products available. Just remember that cannabis is going to be consumed, so don't use anything that is intended for ornamental plants. Use only products designed for vegetable gardening, and preferably organic.

Total Dissolved Solids (TDS) gauges the different blends for vegetative or bloom growth. Remember that tap water already has minerals in it. The harder the water, the greater its mineral content, so take that into consideration. Use a meter to gauge content; don't just guess. There are a wide variety of products and equipment such as gauges, meters and test kits to use to monitor and adjust all these garden variables. Get reasonably good equipment — UL approved for insurance purposes — and learn how to use it rather than just guessing at what you need to do. Once you find your own "magic recipe" for success, stay by it, track your progress and make adjustments as you go.

If you choose a hydroponic setup, you will need to use reservoirs to contain the mixture and pumps to circulate them. Reservoirs should be aerated with an aquarium air stone. In ebb-and-flow systems, the plants' roots — anchored in rockwool, coco, lava stones, etc. — are flooded and drained several times per day. The medium absorbs and holds water for the plant between feedings. In aeroponics systems, water is sprayed onto the plant roots on a regular schedule. Deep Water Culture systems suspend the root ball in a basket with roots hanging directly into the nutrient solution. Where there's water, there's risk of spillage, leakage and mold, so listen for the sound of trickling water and be ready to clean up quickly and sterilize any affected area. Install a pond-liner sheet in the grow room to contain any flooding.

Hydroponic systems replace soil with a growing medium like these clay pellets to plant in. Nutrient-rich water flows around the roots for efficient feeding. Credit: D-Kuru

Lamps, pumps, timers, fans, dehumidifiers, filtration systems: All this equipment burns a lot of power, so it is imperative that you have your electrical system professionally installed and up to code. Make sure you have sufficient breaker capacity for the expected electrical draw of all your lamps, and add at least another 25 percent capacity to handle all the other equipment — your fans, pumps, air conditioning, de-humidifiers, etc. If you add a 1.5 amp breaker line for each 1,000-watt lamp, you should have a reasonable buffer of safety.

Most garden fires don't start from faulty wiring, however. Electrical overloads usually just trip the breaker but have been known to blow out whole transformers. An ethanol-damp cleaning cloth left sitting on a lamp hood when the lights are off can catch fire when the lamps turn on and hit their flash point. Lamp heat can also burn or over-dry the plants, so many growers have switched to using 600-watt or even 400-watt lamps that use less energy and can be placed closer to the plants, which in turn helps keep the room cooler and conserves energy, to boot. This is also an advantage of CFL (compact fluorescent lighting) and LED (light emitting diode) lamps: less risk of fire, less electrical draw and fewer heat-control issues.

## PESTS & PEST CONTROLS

The best advice is never to have pest problems. Practice good garden hygiene. If you are taking a commercial product to market, consider the merits of hairnets, sterile entries, gloves and using tongs or chop sticks to touch the harvested buds. The federal EPA offered guidelines to the Colorado Department of Agriculture Division of Plant Industry on May 19, 2015, regarding the "Special Local Needs Registration for pesticide uses for legal marijuana production." Likewise, one can follow the standards set in the American Herbal Pharmacopoeia monograph *Cannabis Inflorescence.*

No matter where you grow, your garden will encounter adversity. It may be mold and fungi, it may be chemical imbalances, it may be insects, it may be cops, it may be deer — but something will come along to challenge your growing skills. The nature of these pests may vary from indoors to outdoors. Outdoors, you encounter more pests like rats, rabbits, deer, powdery mildew, botrytis, caterpillars, snails and slugs; whatever happens to wander by your garden to take a taste. Indoors, you encounter more spider mites, aphids, white flies, mealy bugs; anything that can catch a ride into your garden, invade and set up a colony. The easiest defense is to prevent infestations by sealing the room and keeping it clean or handle them while the damage can be controlled swiftly and innocuously, like adding sticky traps for white flies. Another trick is to grow other plants that the pest likes better, such as growing eggplant near your cannabis plants. The mites will move over to the eggplant where you can kill them without contaminating the cannabis crop (but don't eat the eggplant!).

Keep several quarantine areas for new or suspect plants until you feel they are safe enough to mingle with your healthy ones. Each set of problem plants should be separated for two to three weeks to see what develops. When plants are infested with insects, you need to anticipate one or two generations of pests to come. A clone plant with just a trace of a virus can ultimately infect the entire garden, leaving you only with the choice to destroy the entire crop, clean everything up and start over — a devastating waste of time and money. The next crop should be started from seed rather than trying to "rescue" plants by taking cuttings because the virus often infects the cell tissue of the parent plant. Starting from seed in a clean environment gives you a fresh start but it also means you have to sex and evaluate all the plant stock again and you might never get back the genetics of the infected plants.

While you're at it, keep your pets out of the grow room. Those hairs get everywhere and a stray hair at the wrong time could cost you a customer.

## THE GARDENER'S PESTICIDE DILEMMA

Once you encounter a problem, resolve the issue without toxifying the crop. That means trying to be organic, but at least use no systemic pesticides that are designed for ornamental plants. Don't use rodenticides that will later poison predators like hawks and owls that feed on mice and rats and become victims of second-hand poisons.

Use only pesticides and fungicides that are rated for consumables and break down quickly; keep anything even potentially toxic away from plants during the flowering phase. Pyrethrum and neem oil are two natural pest controls that are safe, particularly early on during the growth process. Using diatomaceous earth around the roots to stop nematodes or away from the plants to stop ants from getting to them will prevent aphids without affecting the plants at all. Use hydrogen peroxide or even chlorine to clean an area free of eggs and give your garden a fresh start, and then repeat two weeks later. Use biological controls like ladybugs and praying mantis to patrol your garden; just make sure not to harvest them along with the crop.

Diatomaceous earth, an organic white powder made from ground shells, repels and kills many species of insects. Spread it around the base of your plants but wear a particle mask when applying so as to not inhale the powder.

Spider mites aren't just ugly – they are arachnids that can destroy an entire crop in a few days. Webbing spun around the flowers is a telltale sign that you have mites to kill. Credit: Charles Lamb

When the plants are vegetative, you have more freedom because people don't consume the leaves unless you use them for resin to make butter or extracts. But when the plant is flowering, it is difficult to get rid of pests without ruining the crop; any toxins may continue to contaminate the product after it has been prepared. That brings up both health and liability issues, so if you recently sprayed anything, don't try to use or salvage it. Just let it go. You won't want to poison yourself, and it's the price of doing business in the consumer market.

## ORGANIC HORTICULTURE

One of the greatest benefits of cannabis policy reform has been to introduce consumers, and especially the patients who need them most, to organically grown strains and products. Given the choice, who would opt for cannabis covered in noxious chemicals? And legalizing states have already begun to solve a longstanding problem by tracking cannabis products back to the source, so consumers can be assured that products labeled as organic actually meet the standard.

These growers are making their own compost to recycle their old material and reclaim nutrients for the soil.

So growers should choose organic inputs whenever possible, but true organic gardening goes deeper than that — literally. Cultivate a relationship with your soil. Make it a welcoming home for mycelium and beneficial worms. Let your cannabis plants be part of a virtuous cycle; let their deep-tapping roots help to aerate the soil and let it breathe. Work holistically to improve your soil, and watch your soil improve your yields.

Beyond organics, some cutting-edge growers are taking clean cannabis to the next level with veganic techniques, pioneered by cultivation and breeding expert Kyle Kushman. The veganic approach eliminates not only toxic chemicals but also all animal products from the garden.

## ETHICAL ISSUES AS A GROWER

The pesticides question raises a key point. As we move into an age where indoor cultivation of cannabis is the norm rather than the exception, there are important ethical issues to reconcile. Using natural sunlight to the fullest extent possible is good environmental and financial sense. Likewise, the plan for any business should be its sustainability over the long run. Using banks of low-end HID lamps immediately increases both yields and profits, but what are the implications for the future, when electric rates rise? The higher upfront cost of using LED or even CFL lighting pays itself back through long-term utility savings and carbon offset. So does using alternatives to air conditioning when possible, like running indoor gardens during cooler times of year. Taking steps to use closed-loop systems that recycle water and other resources saves on your utility bills and waste output, but you have to be careful not to inadvertently spread fungi or insects. Every choice you make shapes the future of cannabis.

## ECONOMIES OF SCALE

When planning a garden, there are factors that come into play based on the size of the operation. One thing is how conspicuous the location will be. A small grow can fit inside a warehouse space, residence or garage.

For example, there are numerous advantages to setting up an indoor garden in a warehouse in an industrial area over doing it in a house. Most home electrical wiring can't handle an indoor grow with more than about 4 HID lamps. If you use three of those for flowering, get the standard pound of bud yield per 1,000 watts of HID light in four successful harvests per year, you end up with 12 pounds of bud from the residence. The more electricity you use, the higher tier your utility bill will charge you at, so the power bill goes up astronomically as you add lamps and equipment. In an industrial area, the more power you use, the lower the cost per unit and you also start with a greater capacity of size and infrastructure.

Likewise, various compromises arise on whether a crop is worth saving or should be discarded and you start over. A smaller infested crop might be worth the effort, but a larger crop poses a risk to future grows, so cutting your losses to clean the facility and start anew might be safer and more practical. Questions will arise about drainage and ventilation capacity, how many gardeners you can afford to pay to work the crop, hand trimmers or machine, etc. So before jumping in and writing a lot of checks, make sure to consider all these little adjustments, how they affect one another and where you expect to end up in the process.

## HOW LARGE OF A COMMERCIAL GROW?

How much of your time — daily, weekly, monthly — will it take to grow one pound, according to the method you choose? If you don't already know, guess conservatively and assume the worst — especially on your first grow. Then, once you start, track your time and expenses. List all the ways that your operation taxes you. Note every time you had to drop what you were doing to put out a (hopefully metaphorical) fire. Compare the time spent on these tasks to the total amount of time you're interested in working in a day, week or month. Did that pound take up 20 percent of your time? Thirty percent?

Be realistic. If you'd like to ever take a vacation or hang out with friends or go on an occasional date (as many an overcommitted grower only wishes he could do), budget that time. Make sure you are not intruding on it with your commitments to your grow. And unless you're one of those medical marvels who never gets sick, make sure to give yourself an extra 10 to 15 percent for convalescence. To do otherwise is to assume everything will go perfectly.

Suppose you discover you can personally grow and process about 10 quality pounds consistently per cycle. Does this mean you'll be subject to a 10-pound limit for your whole career? Not necessarily, but be warned that growing past your personal limits adds a whole new layer of complication. There are three ways you can go from this point, but each approach should be undertaken with caution.

Many successful solo growers go on to seek out a growing partner who can help them expand. It's a great idea in principle — your partner can watch the grow while you run errands or take a break, and you'll have an extra set of hands around — but watch out. Your new partner may be good at making himself look like a better grower than he actually is, or maybe he is genuinely skilled but has an approach that is incompatible with yours. At worst, maybe he's just a con artist looking for his next mark. To make a growing partnership work, you must have a partner you know is trustworthy, and be prepared to clearly communicate and get it in writing. Otherwise, pass.

Growers with a bit more liquid capital may hire gardeners or trimmers, especially during the busy harvest season. This is often a necessary step, but it has the same pitfalls as finding a partner: Are they trustworthy? Are they competent? Can you afford them? Are they experienced? Growers who hire novice workers usually spend more time training and managing their help than on their grow. Successful employers screen their farmworkers for training and experience. Employee placement services and job fairs are beginning to fill that need.

A third option that has recently grown in popularity is the hiring of a consulting service. Some firms offer full build-out and turnkey operation services that include their own pre-screened staff with plenty of experience working together. This can be a wonderfully convenient option for a grower looking to expand, but it comes with a big caveat: it can be difficult to discern the difference between consultants who can deliver on their promises from con artists or incompetents with over-inflated egos, and the few consultants who can actually deliver on their promises tend to be very expensive. Proceed with caution.

## FINDING YOUR NICHE

Once federal legalization occurs, large corporations will try to crowd all other growers into niche markets. Luckily, in the cannabis economy, niche markets are king. The kind of artisanal quality that most heavy cannabis users prefer cannot easily be produced en masse on giant plantations and processed by giant machines. This very group consumes the largest part of the cannabis market supply, so your business interests largely lie within the realm of their interests and those of the wholesale and retail and merchant. Consequently, most of the cannabis market is in one kind of niche or another.

What's your niche? There are so many ways to grow cannabis that the possibilities are limitless. Next, let's look at what you can do with all that good bud you will be growing.

# 13 HARVESTING & PROCESSING

In this chapter we will presume you already have a crop of cannabis ready to cut down and harvest, or access to some raw material. But if you have never grown cannabis before, be ready for a letdown.

When the plants are growing lush and heavy on the stalk, it looks like you will have a lot of bud material. But about 75 percent of that mass is water that will evaporate. How much of the remaining 25 percent is bud and how much is leaf or stalk depends on how big the plant got and how much leaf

Before harvest.

Hanging to dry.

was left on it at the end of the growing season. It could be more than half or less than a third of the dry plant mass, depending on its condition before you harvest it. It's not that your yield won't be good, because with any luck, it will be. It's just that you will watch it dwindle down and down in size as you process it.

Cure in a jar before the final dry and packaging.

Track your weights throughout the production process. For better reliability, have a separate scale for larger weights (pounds, kilos) than for smaller ones (eighth ounces, grams) and be aware of the standard deviation. Ambient humidity in the air can cause bud to fluctuate in weight.

All the various steps you take along the way will determine your final garden output. Eliminate problem plants as you go along. Spread the canopy branches evenly in the light and use judicious pruning as harvest nears. You are waiting for the approximate flowering cycle for the plant's genetics in the specific context of the garden's growing environment to make your move. Any amendments added to the plants during their vegetative and early flowering cycles should stop for the

Resin / trichome frosting.

last week of the grow to flush the plant out with good, pH-balanced water. Stop feeding the roots. That will clean the product of any unwanted chemicals and improve its taste.

As your expected maturity date approaches, you need to start watching the flower stigmae, those little white hairs popping out of the calyx, crowding with crystals. You want to see them turn amber; at least 60 percent of the hairs should be noticeably more warm in color. The flowers become more dense and sticky to the touch (but don't touch them too much or you'll damage the resin glands). Trichomes appear along the inner part of the small leaves; they increase in brightness and density. Use a magnifying tool to watch the trichomes develop. Use a microscope to examine resin and plant samples for impurities. You will see them start as small reflective pools, then shape into protruding spikes. Once those spikes get heads on top that make them look like little mushrooms, they become known as capitate trichomes. When they turn milky and start to collapse, you need to make the harvest call.

Optimum cannabinoid balance is a point of debate, but if maximum THC content is the desired outcome, you'll want to harvest when about half of the trichomes have turned. After that point, the plant gets diminishing returns and begins to lose potency in the bright light.

Now it's time to cut the plants. If it is a big plant, it is often best to cut it at the base, then trim off the branches along the trunk stalk. The side branches can be cut into smaller lengths or hung to dry. Some people pull the leaves and manicure the flowers while they're still moist, then dry the buds on screens and use the trim for water hash. Others remove only the large stalk and fan leaves, then hang the plant branches to dry first, usually hanging from a line, and then do the manicure. In either case, the drying should be done in relatively cool, well-circulated air. Use a dehumidifier if necessary.

Long branches hanging to dry.

After the bud has dried for a week or 10 days, it is closed up again, normally in a plastic bag or jar, to cure. This does two things: first, it brings out the terpenes and makes the bud more aromatic and flavorful; second, it wicks the moisture out of the heart of the bud and allows the drying process to finish. After another four to seven days, the container is opened up again for a few

Hanging smaller branches on slide out racks saves space but requires good air circulation.

more days to "burp" out the moisture and find its perfect ambient humidity. If you fail to cure the bud, it is less flavorful, less potent and more susceptible to mold. That's right, even after all this, your product is still at risk of molding or spoiling, so it should be kept dry and in the dark to preserve its potency as well as its usability.

A piece of stalk is kept but only to retain the bud's form (a longer bud has more stalk), and leaf and sugarleaf are manicured away using scissors.

Paper sacks let bud breathe and reach ambient moisture levels.

The flowers and even some of the leaf material can be further processed into edibles, resin, tinctures, topicals and extracts, which we will discuss later. For our gardening purposes, once the garden has been cut down, cultivation has ended, but the process of drying buds and preparing the crop continues.

Once the cure is finished, label and store in jars in a cool, dark location.

Some bud will be immediately consumed and some will be stored. The bud can be stored in a dark, cool place for five to 10 years and still have some potency, but for optimum terpenes and freshness, you have up to a year.

This begs the question of what are you going to do with the leftover shake and trim that you can't use for bud because it's too broken up or contains too much low-grade material. The samples below show mixed shake and fan leaf above, shake with undeveloped "air bud" or "trash bud" on the lower left, and ground-up shake mixed with trim that is prepared for extraction.

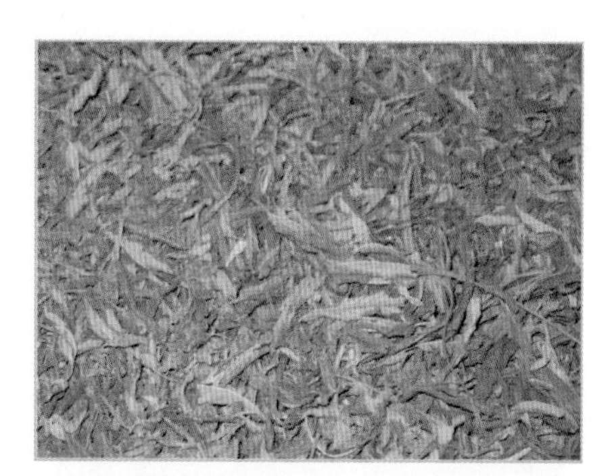

This shake is mostly leaf.

Stalk has traces of resin but any debris must be removed before further processing.

Better quality trim has sugar leaf and bits of "air bud" that had not fully developed.

Grind the material before making extracts or edibles.

## EDIBLES

Edibles are occasionally made in-house by retailers, and this can be another great opportunity. Not every retailer has a culinary flair or a commercial kitchen, but they can still prepare extracted trichomes from shake into butter or coconut oil for their own use or for customers unsure about how to make butter for themselves. A double-boiler process is usually used for cannabutter collection. Olive oil often just involves a warm soak of the cannabis, then through a strainer to clarify the oil. Olive oil and butter are among the best forms to retaining those flavorful terpenes. It is also possible to add extracts to foods as long as the consistency works.

### EVERY NEWBIE NEEDS TO KNOW:

Cannabutter potency varies according to how much cannabis you use, how potent the cannabis is, and how well the extraction process works. Typical cannabis-to-butter conversions: 1 pound low-grade shake per pound of butter; ¼ pound of better grade shake or "trim" per pound of butter; 1 ounce of bud or top-grade trim per pound of butter. This means that for each pound of cannabutter you use in your recipe, you will get about 100 mild doses or 50 strong doses. Try an eighth-ounce kief or a gram or two of oil per pound of butter, and adjust as needed.

Edibles from the Califorganics.

## TOPICALS

Topicals use a similar process to edibles. The resin is coaxed out into a cream, lanolin, isopropyl alcohol, coco butter, oils, arnica, etc., and mixed with other treatments into a topical application, like sprays, creams, salves, liniments and compresses. While cannabinoids may reduce the irritation caused by DMSO, that transdermal compound is seldom used with cannabis topicals. Likewise, extracts can be blended into the other ingredients to good effect.

**EVERY INDUSTRY NEWBIE NEEDS TO KNOW:**

To prepare mildly medicated "cannabutter," use 1 ounce of cannabis shake to 1 pound of butter. Remember that the butter and other food products have a shelf life, so consume products while fresh.

## TINCTURES

Tinctures play the middle ground between smoking and edible. They are typically made by soaking cannabis in warmed glycerine or liquor-type alcohol and then straining. The tincture can be mixed with flavorful or medicinal substances and is placed into a bottle often with a dropper. The consumer periodically takes a drop to a squirt sublingually as needed or desired. It has a quick onset and the psychotropic effect is distinctly different than when smoked. Adding extracts into other types of tincture can be tricky, and time and temperature are often keys to how well the cannabinoids are distributed if you use that method.

## CANNABIS CONCENTRATES

The compressed resin extract called hashish is the most ancient and well-known of the cannabis extraction technologies. For thousands of years consumers have made hashish by rubbing the resin of the flowers off into their hands for *charas*, but today the most common method to make concentrates is water.

After making dry ice hash, a film of kief remains on the extractor.

## MECHANICAL

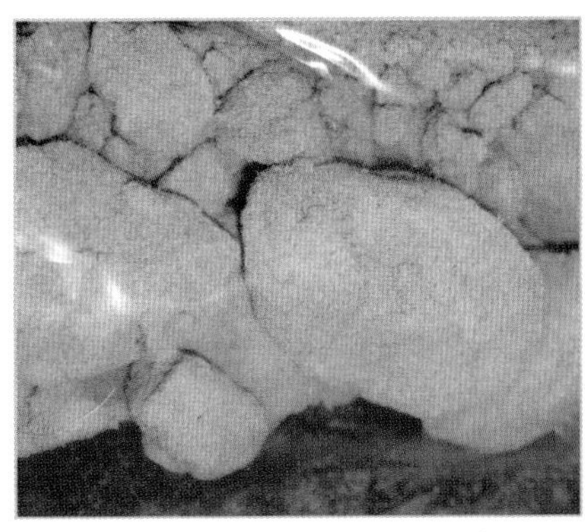

Kief inside a plastic bag. Credit: Chris Conrad

Extracting cannabis resin mechanically can as simple as sifting it off — known as kief — over a 40-120 micron screen. People even screen their own concentrate at the bottom of their personal grinders. The trichome powder collected can be a pale green or a glistening golden brown, and is quite potent. Resin can also be extracted using dry ice with a Kanga Can, water hash bag or other system. This freezes the fresh or dried plant matter and breaks off the trichomes as it is agitated over a screen.

## WATER & ICE HASH

Water hash came on strong in the 1990s, but to compete with solvent extracts, hashish extractors have stepped up their game. A new level of hashish, known as solventless extracts, are gaining traction even in the highly competitive dabbing market (see below) and, newbie beware, the finest of this new breed of hashish can go toe-to-toe with any solvent extract on the market. Tread lightly.

To make the purest, cleanest solventless extract, you'll have to master five variables: stock, temperature, agitation, filtration and time. The best plant stock is densely coated with resin. Most extractors use the crystal-covered sugarleaf trim that is a byproduct of manicuring bud, but the highest-quality extract is run from the bud itself. You can buy kits for making ice hash.

Keep the ice water as close to freezing point as possible. As the water warms, it will extract more chlorophyll and other impurities that will degrade the taste and potency of the hashish. Some agitation is required to coax the delicate trichomes of resin. You can use a kitchen beater, but in general, the gentler you agitate the water, the better. In years past the most popular method was to use a converted clothes washer, but these days the best artisanal hashish comes from water that has been gently stirred, often by hand. Since it is done in water, there's no need to wait for the plant matter to dry, and there is a growing recognition that the most flavorful solventless comes as soon after harvest as possible.

To filter solventless extract, use a fine screen between 25 and 150 microns. A coarser screen will admit particulate and reduce the purity of the end product; a finer screen could filter out the trichomes and defeat the whole purpose.

## SUPERCRITICAL CO2

As potent as solvent extracts can be, perhaps the most powerful cannabis treatments are produced using supercritical carbon dioxide (CO2) in a closed-loop solventless system.

Closed-loop supercritical extractors can be both expensive and bulky (see below), but in the quality of medicine they produce, they are well worth the investment. The technology to produce CO2 extract is so advanced, you can get high just *thinking* about it. The cannabis plant material is placed in a chamber where it is blasted with carbon dioxide, which is kept at a precise combination of temperatures and high-pressure settings so it exists simultaneously as a solid, liquid and gas — a supercritical state. The CO2 breaks off the plant's potent cannabinoids but leaves behind the rest of the plant material, yielding an extract so free of impurities that you could read this book through it easily (go ahead, try it — we'll wait). The closed loop means that the CO2 is converted back to a gas, drained out of the chamber, and then reused.

Closed loop supercritical CO2 extractor.

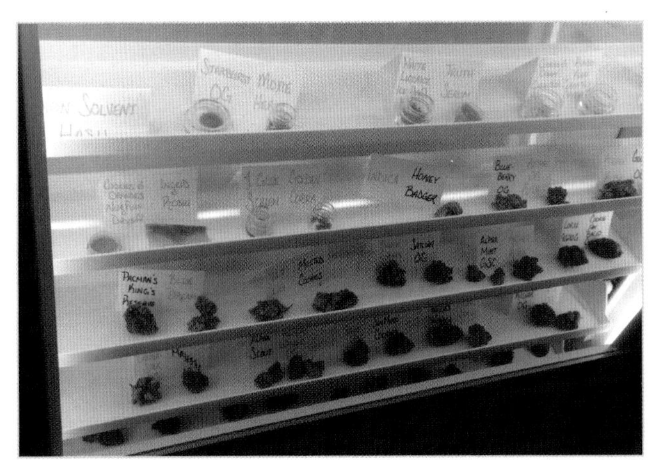

A selection of bud and extracts from a High Times Cup.

## SOLVENT EXTRACTS

WARNING: Most states, even ones that have legalized marijuana, continue to regard some of the techniques described in this section as criminal offenses — even felonies. Do not attempt any of the techniques in this section without consulting an attorney competent in the laws concerning solvent extraction in your state.

Liquor grade, high-proof grain alcohol can be used to dissolve the resin and remove it in suspension from the plant matter, which is filtered out. Sunlight or low heats are used to evaporate the alcohol and leave the oil extract, or the suspension can be used as it is with the alcohol base. This process is pretty harmless, maybe like the risk of serving a flambé dessert.

Golden butane honey oil is praised far and wide for being a very concentrated, very potent medicine. For some patients, it is the best thing ever, but blasting BHO has caused many severe injuries, even among people who thought they knew what they were doing. Even in the unlikely event that you may legally do this, do NOT do it on your own. PVC pipes and coffee filters are not good engineering. Careless solvent extraction can lead to explosions big enough to blow up a house, leading to injuries and potentially death. To engage in this process you need the right facilities and comprehensive training on the safe extraction of cannabis resin using solvents.

The greatest risk of explosion from butane extraction stems from escaped gas that, being heavier than air, can pool along the floor or behind appliances, like refrigerators or ovens, and flash with the ignition spark. That's why closed-loop extraction systems are a must. For a few thousand dollars, extractors can invest in a high-quality, leak-proof design that recycles butane

### What Is Oaksterdam University?

For the most dedicated cannapreneur, there's no substitute for Oaksterdam University, the nation's leading brand in cannabis education. Founded in 2007 in uptown Oakland, California, by activist and entrepreneur Richard Lee, Oaksterdam has since grown as rapidly as the prodigious plant that inspires it. Hands-on classes in horticulture, law, history and rhetoric are still taught at its Oakland location, but the curriculum has gone on the road to Colorado, Massachusetts, Nevada and Washington, D.C., among other locations. For a cannabis college that's second to none, see oaksterdamuniversity.com.

after each use (up to about a dozen times) and drastically reduces the risk of escaped solvent.

Some solvent extractors prefer 99 percent isopropyl alcohol or even naptha; these solvents produce a thick, syrupy substance known as Full Extract Cannabis Oil (FECO), Rick Simpson Oil (RSO) or Phoenix Tears. The resulting extract can be powerful medicine, but remember to work the oil with a blade to build texture and release any gas trapped in pockets in the end product. Rinse with water to flow away any residues from these solvents. The purity of the end product will affect texture. If more plant waxes are retained, the material would be wax, peanut butter or honeycomb. There is disagreement about how whether it is better to eliminate all terpenes other than cannabinoids (especially THC) or better to retain as many terpenes as possible. Our guess would be that you want to keep most of the terpenes. Due to the toxicity of and fire danger posed by the solvents used, we don't recommend that any newbie attempt these techniques.

Solvents create bubbles as they evaporate and leave the resin.

## Hand Trim Or Mechanical Trimmer?

The manicure can be done by machines that tumble the buds over blades or it can be done by hand. There are a number of reasons why either method might be preferred in a given situation. For example, if there is a large amount of bud to manicure, using trim machines will make shorter work of it. Hand trimmers are more selective and better for smaller harvests of artisan quality buds. It takes fewer personnel to run a trim machine than to do hand trimming. Machines tend to break off more trichomes than does the personal touch, but if the cannabis is good enough, a difference of a couple percentage points in potency is not going to be a problem. If a consistent generic quality of manicure is desired, the machines output is adequate, but to a talented trimmer, a cannabis bud is a work of art.

# 14 PROPAGATION & BREEDING

Seeds are the traditional way to propagate cannabis, but they have to be fully matured in order to sprout, which means they need to be dark and often spotted. Green or cracked seeds will not sprout.

Growing from seeds has the distinct disadvantage of producing, on average, 50 percent male plants and 50 percent female. Unless you plan to grow to produce seeds, the male plants are worthless as marijuana — so time and resources used growing them are wasted. Feminized seed lines have a much higher percentage of females to males, up to 100 percent female offspring. The seeds are usually soaked overnight and sometimes sprouted in a paper towel before being popped into a growing medium such as soil or a grodan or rockwool cube. These are kept moist for a week or so, by which time the sprouts should be coming up. To help the starts grow and keep them from drying out, they can be kept in a warm location with a cover over them to

keep the moisture in, but once the sprouts start to pop up, you have to remove the cover because they may either mold or hit the lid and begin to grow. You want the stalk to be straight at this point, not crooked.

The first set of leaves are called cotyledons; they are round and don't look like regular cannabis leaves, which have serrated edges and are pointed at the ends. The next pair of leaves are single blades with the typical cannabis shape, followed by a pair of three-bladed leaves, and after that, classic five-bladed leaves. As the plant grows, it continues to add blades in odd-numbered sets: Seven, nine, 11, even 13 blades on a single leaf.

A female plant started from seed from a stable seed line or a hybrid plant can be used for flower, pollinated for seed or serve as a mother plant from which to make cuttings that may take root and become clones. Those clones will always be female, and can be grown into mother plants in turn to make still more clones, and so on for generations. This process is usually done in a greenhouse or indoor nursery. Cloned plants lack the taproots and cotyledon leaves that sprout from seeds.

Roots penetrate the rockwool.

A cutting grows into a clone.

Clones labeled, started in rockwool plugs and inserted into soil.

At one time, making clones was an arduous task with a low success rate. It required identifying nodes and making precise cuts with a scalpel.

> **A newbie's lexicon (glossary)**
>
> *Cotyledon leaves* are the first, rounded leaves to appear on a seedling.

Today there are rooting compounds available that make the process easier, with a much higher success rate, but it still doesn't hurt to get a good slice below a healthy node when making those cuttings. Clones can be grown for bud, pollinated to produce seed or maintained for mother plants.

Like seedlings, baby clones should be kept warm and in a protected, moist environment. There are special starter trays that make this process simple. If possible, start growing more plants than you expect to harvest as a hedge against plant loss or, if the plants all survive, to select only the best plants to allow to grow to fruition. As your survival rate improves, you can make fewer starts at a time. Label the clones by strain to keep tabs.

Cloning trays are covered and labeled by strain and date to keep track.

## BREEDING SEED LINES, MAKING HYBRIDS

Breeding cannabis is a labor-intensive and time-consuming process that involves growing a lot of plants. The goal is to start a new seed line or cultivar that has some unique character that sets it apart from other varieties of cannabis. For decades now, the main driving force has been to create and market cultivars that produce the most THC, to the point where other cannabinoids, and particularly CBD varieties, were being kept from the gene pool.

The genetics have either dominant or recessive traits, so typically the goal is to create a stable seed line showing a desirable trait. If it is a dominant trait, breeding for it is easy to achieve but it will likely be a common characteristic for other varieties as well, so it might not classify as a unique strain at all. Breeding for recessive traits or certain aesthetics such as fragrance or cannabinoid profile requires that the seed stock consistently show the same characteristics for several generations, which makes it a stable seed line.

What are you looking for that makes these genetics particularly desirable? Plants can be selected for their fragrance: Is it citric? Skunky? Fruity? They can be selected for their resin production. Does the bud look frosty? Are the hash glands (trichomes) sparkly and close together? The more resin, the more cannabinoids.

Sometimes it is the color that stands out. Does it have a purple or bluish cast to the flowers and upper leaves? A bit golden perhaps? Maybe the hairs give the bud a reddish amber tone? Are the bud formations especially dense and tight, or are they light and airy? Do the flowers ripen quickly or do they take a longer time to finish? How does the bud taste when smoked, and does it have the desired medical or psychotropic effect? Is it mold resistant? Is it drought resistant? As you select for these and other specific sets of characteristics, called phenotypes, you have to match up the plants to breed and cross-pollinate. If the characteristics become stable and consistent over multiple generations, you may have a new genotype (or you may have duplicated someone else's plant line). If the desired characteristics are prominent and consistent for the first generation but can't be maintained for

Seedlings.

future generations, you have an unstable hybrid strain.

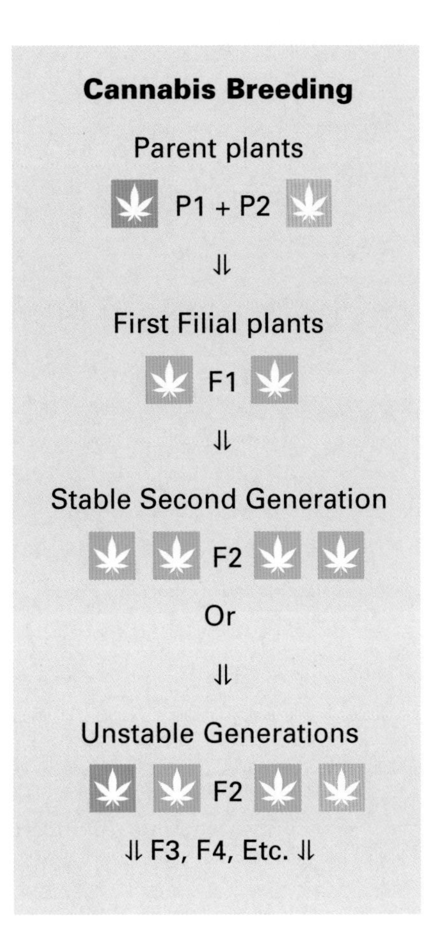


Why would you breed a hybrid like that? Because the breeder will identify two strains with complementary characteristics that produce reliable offspring; say an indica variety crossed with a sativa variety. Indicas are compact with a strong indica effect and finish well indoors. A sativa strain has a sweet flavor and an ethereal effect but matures slowly over the long outdoor flowering season. Combining them would give you a potent and flavorful plant that can handle being grown indoors or outdoors and has a medium-length maturation process. The next generations can be inconsistent in effect, may have leggy plants or squat ones, some taste sweet while others are sour or ripen erratically, needing constant attention. It doesn't matter if the resulting seed line is stable or not, because only the first generation seeds are meant to be grown. From the seed producer's point of view, this is an advantage because a grower can't simply produce their own seed stock from year to year; they need to buy the breeder's seeds (or their clones) in order to have access to the parental genetics.

Cannabis is an annual plant, but a farmer growing indoors can harvest three to six generations of fast-growing varieties in a single year. Using indoor lighting cycles, it is possible to keep a female plant or a male plant alive for years. Pollen, too, can be stored for years to fertilize females, which allows back-breeding successive generations of cannabis plants. These projects can get pretty intense and involve many thousands of plants to be successful, but the rewards can be amazing.

## SOURCING GENETIC MATERIAL

Before you get too involved in this process, you should decide whether you have the patience and wherewithal to devise a new strain of cannabis, and how you are going to promote and market the seeds. Also, remember that the genetics you may adopt to fashion a new seed line have come from someplace or someone else and, especially as DNA tracking of the plant becomes more detailed and refined, another breeder might make a claim on your hard work. The heritage strains contained both THC and CBD, but as consecutive generations were hybridized, CBD was bred out of marijuana strains.

## HERITAGE CANNABIS GENETICS

- Mexican: Sativa, sweet: Original 1960s and '70s seed used for plant selection and breeding.
- Columbian: Sativa, sweet: One of the first upscale imports, grown and processed more carefully; more potent and fewer seeds than Mexican.
- Afghani: Indica, pungent: Among the first seeds collected by American and European breeders traveling the historic spice route; Hindu Kush and Afghani are foundations of indica lines.
- Thai: Sativa, sweet: Bud tied to bamboo by thread; few seeds.
- African: Hybrids, more Indica in the North, more sativa in the South: Durban Poison is sativa dominant; Ghana strains more indica; Lebanese and Moroccan primarily used for hash.
- Jamaican: Mostly sativa hybrid: The island was a crossroads for Hindi, African and American cannabists who traded seeds; in Jamaica, it's still known as ganja, its Hindu name, while top grade is called Kali.

## SUBSEQUENT HYBRIDIZED GENERATIONS

- Original Haze, nearly pure sativa, sweet fragrance; Mexican, Colombian, Thai and a sativa from India. Four phenotypes: Gold, Silver, Purple and Blue. Resinous, hard to grow, very long flowering cycle.
- Northern Lights, nearly pure indica, dark green extreme Afghani indica with a small Thai cross: short plant, dense flower ideal for indoors.

- Skunk hybrid, mostly sativa Columbian x [Afghani x Acapulco Gold Mexican] named for its pungent odor and potent effect.
- Jack Herer hybrid by Sensi Seed, from the three classic hybrid strains and named after the late author and cannabis activist.

Theweedblog.com listed the 25 most popular strains of 2014 as Blue Dream, Sour Diesel, Girl Scout Cookies, Green Crack, Jack Herer, SFV OG, Headband, Fire OG, White Widow, Super Lemon Haze, Bubba Kush, OG Kush, Durban Poison, Tahoe OG, Master Kush, Skywalker OG, Cherry Pie, Blackberry Kush, Trainwreck, Purple Trainwreck, Pineapple Express, Super Silver Haze, Grape Ape, Purple Kush and Purple Haze, but there are scores more names and everybody has their own preferences — like Sage or one of the Diesel varieties. A lot of strains come in white, silver, blue, purple, red, green and sometimes blonde versions.

After decades of growing for the strongest THC content, a number of CBD-dominant strains have emerged, such as Harlequin, Charlotte's Web, Cannatonic, AC/DC and variations of Tsunami.

## MUTATION SELECTION

Mutation selection is the thing that makes each plant special, which makes breeding so exciting, so let's end up this section by showing a few mutations of cannabis.

Some efforts have been made, with no reliable success to date, to breed marijuana that does not look like cannabis. The main purpose of such a plant would be to disguise the crop so that it would not be detectable by law enforcement. Of the three variants below, none have been stabilized. The plant with a triple-branching pattern still looks like marijuana but could have a larger yield. Likewise, the lateral flowering may look like a caterpillar on top of a flattened cannabis plant but the leaves still say weed. The third one has triple branching and lobed leaves rather than pointed; since the leaves differ from the iconic marijuana leaf known around the world, it is the best disguise of the three. There are also efforts to breed odorless cannabis, but given the importance of the terpenes to the overall experience, one can only wonder: if a cannabis

**EVERY INDUSTRY NEWBIE NEEDS TO KNOW:**

Keep careful track of the seed lines you use to breed your strains; you'll never know which one will be the perfect combination.

Genetic variations in cannabis.

plant doesn't look or smell like cannabis anymore, will it still have the same effects of marijuana? And if we find a way to breed cannabinoid terpenes into other plants, will they become marijuana? It's hard to say, since hemp doesn't have any psychotropic effects but the government still lists it as marijuana.

## CANNABINOID SPECIALISTS

The Stanley brothers got famous for breeding Charlotte's Web, their special strain high in cannabidiol (CBD) for the treatment of epileptic children. Soon their nonprofit, the Realm of Caring, had to institute a waitlist for desperate parents just because they couldn't produce their oil fast enough. There's plenty of room in the high-CBD market for farmers to grow alongside the Stanleys and others who are filling this niche. Even so, it may be an even better idea to already be setting your sights on the *next* big cannabinoid to make the news — after all, early research into CBC, CBG, CBN and THCV all look promising. Just remember not to overdo it; medical cannabis works by combining all of the cannabinoids together through the entourage effect, so myopically focusing on one cannabinoid to the exclusion of all others is unlikely to produce good medicine.

# 15 ANCILLARY MARKETS

There is more money to be made in selling shovels than in digging for gold, as many a California Gold Rush-era prospector learned to his chagrin. This principle holds every bit as true for the cannabis industry of today as it did for the gold-mining industry of the 19th century.

The ancillary market surrounding cannabis — meaning all of the services and products sold to support the industry, from farm to consumer — brings in far more money than the industry's core of growing and selling bud. The core legal industry, worth somewhere between $1.5 and $2 billion per year (and rapidly growing as more states legalize), certainly offers plenty of big money-making opportunities. But consider the following:

- The U.S. market for consumer hemp products already exceeds $580 million, and as more products become economically feasible to produce with loosening laws, this figure is likely to grow.
- As medical marijuana becomes more pharmaceuticalized (e.g., Sativex, Epidiolex), pharma companies will begin investing more and more into the ancillary market for medical research and development of cannabis-derived treatments. The average cost of developing just one drug through

**EVERY INDUSTRY NEWBIE NEEDS TO KNOW:**

*Marijuana Business Daily* estimates that almost 70 percent of the entire legal cannabis industry is in the ancillary market.

the FDA's process is about $350 million, and cannabis is such a rich source of therapeutic compounds that many hundreds of treatments could conceivably be developed in this way from it. The ancillary medical marijuana R&D market could thus dwarf the core industry all by itself.

- Two of the fastest-growing segments of the legal cannabis industry are the production of edibles and the extraction of concentrates; if present trends continue, the job markets in these two segments can only be expected to continue their phenomenal growth.
- One of the principal drivers of the U.S. economy is the home construction industry, weighing in at a staggering *$82 billion per year*. As prohibitions on industrial hemp finally loosen enough to allow American homeowners the freedom of choice in how to build their homes, more and more will choose the superior quality of hemp-derived construction materials over the inferior quality and unsustainable production of lumber. If even 10 percent of the U.S. housing industry converts to hemp, this alone could exceed the size of the core market.

Hempcrete block construction.
Credit: Oliver DuPort

- Another principal driver of the U.S. economy —domestic consumption of oil — is certain to come to an end in its present form. It's only a question of when. Fortunately, nature has provided a renewable solution: hemp. If even 10 percent of the domestic consumption of 7 billion barrels of oil per year converted to renewable and clean-burning hemp ethanol and biomass, that could be over $10 billion per year set aside for processing hemp into fuel.

So if you don't have a green thumb and the thought of running the complex logistics of a dispensary makes your head spin, take heart: there's plenty of money to be made in the ancillary cannabis market, including hemp. We'll get back to that, but now let's take a look at some exciting marijuana ancillary opportunities.

## CULTIVATION EQUIPMENT & CONSTRUCTION

Depending on whether a grower chooses to cultivate cannabis indoors, in a greenhouse, or out in the sun, the cost of setting up the cultivation site can vary widely. Outdoor cultivation is the least costly to set up, but that doesn't mean it's cheap — outdoor growers still have to pay for irrigation systems, deer fences, landscape grading and terracing, and so on. There is plenty of money to be made in the ancillary market for grow room design, construction, electrical and plumbing installations, equipment and setup, regardless of the cultivation technique a grower may choose.

These covered grow rooms involve ventilation, electrical, plumbing and construction goods and services.

## LIGHTING & LAMPS

While artificial grow lights were formerly the exclusive province of the indoor grower, in the legal cannabis industry their use has expanded to touch on nearly every kind of cultivation. This is due especially to the phenomenal success of light enhancement and deprivation techniques to combine the power of the sun with light-controlled greenhouses and get the best of indoor and outdoor cultivation. Farmers who use such techniques use supplemental artificial lighting to extend their plants' days early in the season and artificial darkness to shorten their days in midsummer, forcing plants to grow longer or to flower early. Entrepreneurs who specialize in outfitting light-deprivation greenhouses with the lighting they need will find there's plenty of money to be made this way in the ancillary cannabis industry.

Lighting equipment options include HID, CFL and LED, seen here.

**EVERY INDUSTRY NEWBIE NEEDS TO KNOW:**

The ancillary market is more competitive than the core, but the financial stakes are higher because most ancillary products and services can be offered on a national market.

Even outdoor growers often use artificial lights. For maximum growth, many outdoor farmers start their plants indoors, artificially lit and heated during the winter so they can develop an extensive root structure by the late spring when they're ready to go in the ground. See "Cultivation."

Of course, the greatest consumers of artificial lighting within the cannabis industry are the indoor growers. Using an average yield ratio of about one pound of finished bud per 1000-watt HID light per flowering cycle, many growers consider it well worth their investment to spend heavily on high-quality lamps. Some indoor cultivation setups cost hundreds of thousands of dollars to build out their lighting and electrical systems.

Remember also that lights eventually begin to dim with continual use, so an enterprising seller of specialty lighting equipment can land themselves some nice repeat business if they understand the needs of growers and how to meet them.

## CLIMATE & ENVIRONMENTAL CONTROL

Aquaponics are booming in the cannabis industry – and out.

To achieve its maximum potential, cannabis needs to grow in the right environment (see "Cultivation"). Given the large gap between the value of medium-quality and top-quality cannabis, smart growers are willing to invest generously in their indoor and greenhouse climate-control systems. To do otherwise is to risk mold or other infection, or a smaller or less potent harvest. Climate-control specialists who have the expertise to always keep a grower's plants "in the zone" thus have a naturally vigorous market from which they can make a very good living.

## NUTRIENTS & PEST CONTROL

Plant nutrient supplies operate by the same rule of economics that govern lights and climate control — growers who skimp on these materials are penny wise but pound foolish. While it is certainly possible to give cannabis plants too great a *quantity* of nutrients and other amendments, in practice there is no limit to the *quality* of nutrients a plant can uptake. As cannabis consumers have grown more conscious about the sources of their medicine, demand for nutrient-rich, organic ingredients has skyrocketed. Will you be one of the many suppliers who meets this demand, profiting handsomely in the process?

The best pest control is preventative, but occasionally a grower forgets this wisdom and needs rapid and urgent assistance to save his crop. In the ancillary market of cannabis pest control and suppression, there is therefore room for at least two kinds of services — preventive care specialists who lower the risk of growers early in the season, or pricey on-call emergency services that are in highest demand right before harvest. Or perhaps an even better model is right around the corner. Will you be the one who growers trust to call when they need help with spider mites?

Preventive pest control is key to growing a beautiful plant like this.

**EVERY INDUSTRY NEWBIE NEEDS TO KNOW:**

Effective organic pest control is in higher demand than ever, as consumers increasingly realize its benefits.

## INSPECTIONS & LAB TESTING

The most important tool for well-educated procurement is a quality test from a reputable laboratory. These tests can alert you to the presence of mold, fungus, pesticides, herbicides and other adulterants that have no place in medicines intended to treat sick patients. Laboratory tests are specific. Adulterant testing is different for heavy metals than it is for molds, and potency testing for THC, CBD and other cannabinoids is a separate process and for each there is a separate charge. Unfortunately, not all THC test reports are to be trusted. Potency varies among various parts of the plant and there are

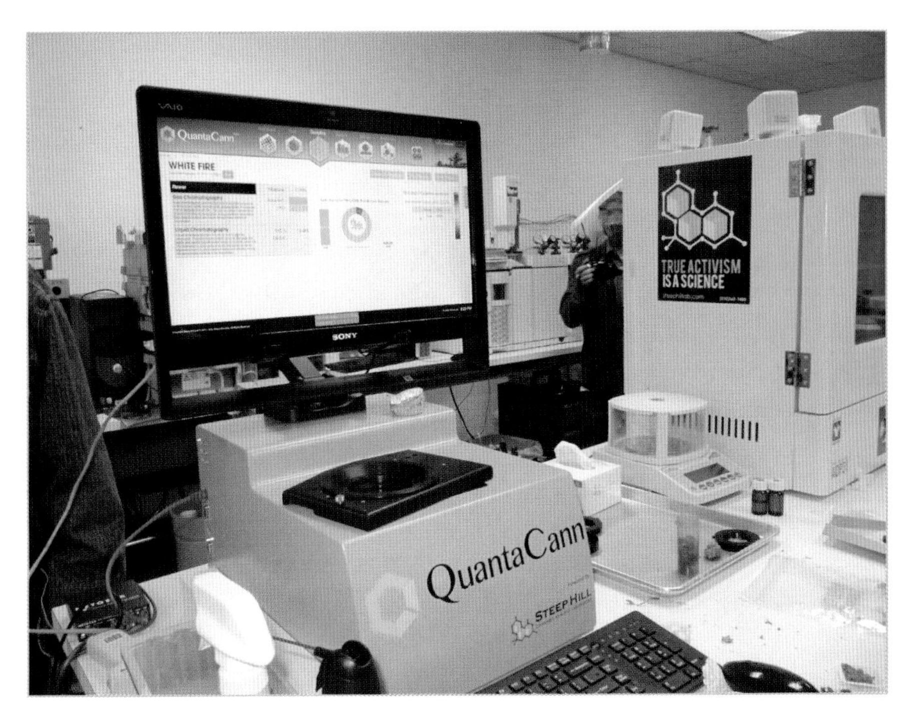


Credit: Chris Conrad

anecdotal reports that some less reputable labs frequently bump up their THC potency numbers to help growers compete in an ever more crowded market. Patients increasingly demand (and states increasingly mandate) data from reputable, licensed third-party labs, making them essential services.

## PROFESSIONAL ANCILLARY SERVICES

To keep the gears of the supply chain running, businesses need a variety of specialized professional services with expertise in the cannabis industry. That means there can be work as a consultant, accountant, legal advisor, investment advisor, lobbyist, conflict resolution mediator or civil attorney to set up corporations and non-profits, and otherwise protect the interests of inventors, investors and consumers. There are also civil service jobs inspecting to ensure kitchens are clean, gardens organic, locations up to code, and consumables safe and well labeled. There are both scientific and marketing job prospects, as well.

**EVERY NEWBIE NEEDS TO KNOW:**

"Less is more"— *Marsha Rosenbaum*

## PATIENT EDUCATION

*New York Times* columnist Maureen Dowd unintentionally opened herself up to public ridicule when she published her embarrassing story of the first time she tried an edible cannabis chocolate bought through Colorado's retail cannabis system. After downing some wine and a couple of nibbles too many from her medicated caramel chocolate, Dowd swooned and "became convinced that I had died and no one was telling me." But while it was almost comical how Dowd attempted to sidestep responsibility for the ways her own naivete contributed to her misfortunes, in one regard at least she has a point: budtenders do need to take care to make sure that consumers — and especially first-time edibles consumers — receive only the best advice on how to consume products to avoid repeating any embarrassing incidents. Make sure your budtenders are well trained on how to educate your newbie consumers, and your business will thrive.

## TERPENE PAIRINGS

Hippocrates, the father of Western medicine, famously admonished his students to "let thy medicine be thy food and thy food be thy medicine." Now, thanks to recent advances into the therapeutic properties of cannabis, researchers are discovering just how wise that teaching was. Indeed, cannabis has more in common with our favorite foods than science once thought. Terpenes comprise one of the most abundant classes of organic compounds found in cannabis resin, and they play a vital role in the entourage effect that makes medical marijuana work. But these wondrous chemicals are not limited to cannabis; they are found in fruits, flowers, saps — practically any vegetable matter that has a pleasant taste or odor caused by naturally occurring terpenes which we instinctually seek out. One of the most surprising developments in cannabis research is the discovery that plant medicines are already all around us.

Inspired by this cutting-edge research, many cannabis entrepreneurs have already begun to take their cannabis edible creations to the next level through terpene pairings — that is, by carefully matching strains of cannabis to choice culinary ingredients to enhance the effects of certain terpenes. (For more information on terpenes commonly found in cannabis, see Chapter 4, "Cannabinoid Chart") Likewise, cannabis-specific terpenes can be extracted as essential oils and used for fragrances and flavoring for things like ice cream or lozenges for all the cannabis taste without the high.

## SOFTWARE DEVELOPMENT

Anticipating the needs of outlets and other sectors of the industry and then creating software programs to meet that need is another path. Viridian Sciences, for example, is offering inventory tracking software from seed to sale reporting, accounting, quality control, grow management, etc.

## FINDING THE RIGHT MEDICINE

Two of the early stars of the cannabis-themed Web, Weedmaps.com and Leafly.com, have thrived by filling the niche of helping cannabis consumers find the products they're looking for — though each website approaches the niche in a different way. Weedmaps, as its name implies, is a map-based app which utilizes the API from Google Maps to locate a curious consumer within their geographic area and identify nearby retail outlets that offer cannabis; the retailers pay monthly fees to be listed, and can pay extra for extra visibility. Weedmaps thus provides an easy way for customers to find cannabusinesses in their area and browse their menus before making a selection. While it isn't known with any precision how much money Weedmaps makes this way, they are apparently doing quite well; their founders have recently ponied up $2 million to support legalization in California.

Leafly fills a similar niche by using a different approach. Instead of a location-centric interface that sorts cannabis outlets by geography, Leafly organizes its information by cannabis strains, leveraging user-generated descriptions to attempt to inform consumers on which strains may be best to treat certain symptoms. Unfortunately for Leafly, this model comes with a major Achilles' heel: the users who fill out the strain descriptions usually aren't doctors, and there's no guarantee that the "Blue Dream" described by one anonymous user bears any kind of resemblance to the "Blue Dream" described by another anonymous user on the other side of the country. Despite this, Leafly was acquired in 2012 by the venture capital group Privateer Holdings.

Each of these sites has demonstrated that a market exists for helping cannabis consumers make better informed decisions, but there's still plenty of room to improve on their models. Instead of categorizing medical options by strain like Leafly, a next-level website might correlate

**EVERY INDUSTRY NEWBIE NEEDS TO KNOW:**

Don't sell out to the first investor group to come knocking. If you've got a good thing going, there will be other bids.

data between symptom relief and chemistry. In a similar vein, whereas Weedmaps has attained a level of success through its strictly local and regional focus, the continuing liberalization of cannabis laws will open up opportunities less limited by geography. After all, Amazon allows consumers to find products from anywhere on Earth; the "Amazon of cannabis" may only be one Congressional vote or presidential signature away.

## CONSUMER ACCOUTERMENTS

Cops call them "drug paraphernalia," and as a result, cannabis consumer accouterments have literally gotten a bad name. But when one peers through the drug war hysteria, they can find incredible opportunities to service the market for helping cannabis aficionados consume their favorite herb. Whereas tobacco is nearly always consumed as cigarettes (more recently, "e-cigs") and alcohol comes in various glasses from beer pints to whiskey sniffers, the options for consuming cannabis dwarf these options in terms of sheer diversity.

## PIPES & WATER PIPES

Pipes are probably the most popular means of consuming cannabis in the U.S., after hand-rolled cigarettes, or joints. With the expanding availability of high-quality vaporizers, this is starting to change; but nevertheless there is still ample opportunity in the market to supply cannabis consumers with useful — and often, stunningly beautiful — pipes.

Ancillary items on sale at Harborside Health Center. Credit: HHC

Although innovative tokers have fashioned plenty of pipes out of metal, wood or even apples and soda cans, the most popular pipes are glass, which is relatively easy to keep clean and can offer the most flavorful experience for the cannabis connoisseur. Options range from small and discreet glass pipes that fit easily in a

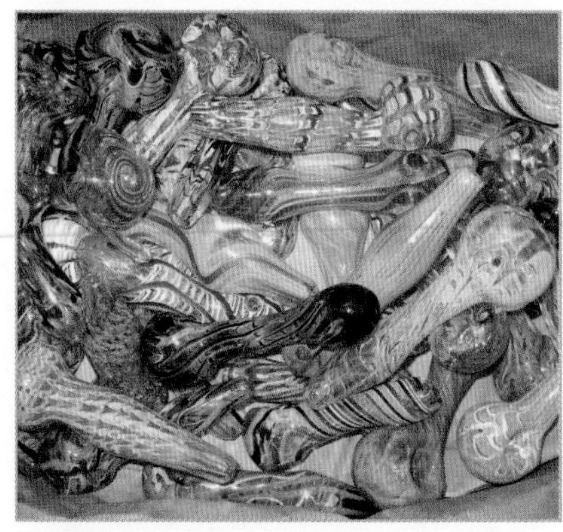

Glass pipes in an array of colors. Credit: Chris Conrad

The legendary Volcano.
Credit: Chris Conrad

pocket to stunningly complex, multi-chamber water pipes that double as astonishing works of art.

Unfortunately, these water pipes and bongs are not as easy on the lungs as on the eyes. While water filtration does cool the smoke and remove some particulates and tar from cannabis smoke, it actually removes a greater proportion of cannabinoids from the smoke than other compounds, so smokers who use water pipes actually breathe in more tar to attain the same level of therapeutic relief. Nonetheless, bongs pack a punch and are popular worldwide.

## HOME VAPORIZERS

An entire ancillary industry is bringing tokers new options for inhaled cannabis: vaporizers. There are pocket vaporizers, hand held, table motels, hookah-style, digital. Vapor Brothers has a whole range of vaporizers. Research the market to choose a vaporizer that is right for you and your budget; it can be a great investment in both your pocketbook and your health.

The one model that has done more than any other to break open the ancillary market for vaporizers is the legendary Volcano. It is expensive, yet countless aficionados swear by it. It uses forced-air convection heating to blow through a small chamber containing cannabis (or any herb) into an inflatable bag to capture vapor that the consumer can inhale from at his or her leisure. The result is a cleaner, better-tasting vapor than smoke that allows the consumer to carefully savor the abundance of volatile terpenes that normally get largely destroyed by the process of combustion. And the Volcano is a worthwhile investment, too; we have enjoyed Volcanos that have been in regular use for over 20 years, and

A vaporizer / bong combo system.

even after all that time they still work great. A nice thing about home vaporizers is they also work with either extracts or herb. But while the Volcano is great, people are always looking for the next big thing, and maybe you are the one who will develop and market it first.

## PORTABLE VAPORIZERS

More recently, the U.S. market has been flooded with a spate of cheap, portable "vape pens" that use small batteries to power a heating element to vaporize cannabis extract. Many come pre-filled and are disposable, others sell better quality pens and then replacement cartridges. Most of these require extracts; the herb models tend to be bulkier and need to be refilled almost every puff. The pens are popular and discreet, but can be hard on the environment and are facing bans in a backlash against tobacco users. Nonetheless, the market is expanding, so that means opportunity.

## DABBING RIGS

Related to vaporizers, dab rigs are more complicated because they need a platform or "nail" that can be heated red hot, a heating device to do so (and often it's a blow torch), usually a cooling chamber that holds water or ice, and enough safety gear to keep people from burning their fingers. What's your design?

## CANNABIS TOURISM & HEALTH RESORTS

A traveling cannabis consumer is often looking for a "420 Friendly" environment and will always choose it over another location. The "Bud and Breakfast" model of B&Bs for cannatourists remains largely untapped. Even an inn having a garden for toking is appreciated, but a line on where to get some cannabis seals the deal. We don't want to treat this too whimsically, however, because medical marijuana health resorts and cannabis for end of life in hospice care are deeply underserved societal needs. It is surprising that there are not more such facilities in legal medical marijuana states — yet.

## PACKAGING, PUBLISHING, PROMOTIONAL & THE MUNCHIES

It used to be just a matter of buying some sealable baggies, but now cannabis is packaged in medicine bottles, jars and all sorts of elaborate and child-proof containers. Every bud that goes out the door is usually in a container and a carrier bag. These can be printed as promotional materials. Likewise, people keep stash containers at home, sometimes a wooden box or a decorative jar. These items often make nice gifts because they can be used for other things than just cannabis. Other promo materials include matches, lighters, posters, magazines, books, printed T-shirts and other clothing, rolling papers and filters. If you go to a Seattle Hempfest and walk around, you will see a thousand great ideas.

The business that gets many a wag for being the toker's weak spot — junk food and the munchies — can certainly target the whimsy of a stoned person looking for something to eat. In the 1970s a variation of kettle corn called "Screaming Yellow Zonkers" covered its packaging in stoner humor and had a great run. Even if it's just a fad, someone makes money and can reinvest in the next big thing.

There is no possible way to list every category of opportunity in the ancillary cannabis market, because the only real limit to the possibilities is your own imagination. What kinds of needs are not being served? Where does your passion lie? Where will you find your niche?

# 16 TIME FOR HEMP

As a newbie, you might be a little confused about the difference between hemp and marijuana. The explanation is simple: Both come from a cannabis plant, but only the marijuana-type strains and products make people high.

The U.S. has a long historical connection to cannabis hemp, as 18th and 19th century farmers were legally required to grow hemp as it was determined to be a necessity for commerce and the national security of the American people. In the early days of the republic, subsidies and tariffs were used

Credit: Aleks

## THC Content In The Plant's Flower

Most modern nations ban the use of THC-rich cannabis for industrial applications. Is that because it makes an inferior product? No, in fact, while low-THC hemp seems to work perfectly well for hundreds of applications, many people argue that breeding out THC has led to selection of plants for the wrong reason and that using high-THC hemp crops could improve the quality and production of hemp materials. We won't know for sure until prohibition has been eradicated, but most would agree that plant breeding should be to maximize its beneficial result rather than to obstruct its development.

to encourage hemp farming. In 1913, the Department of Agriculture sought to increase hemp production and manufactures as a way to save both our family farms and our forests. As recently as 2012, President Obama signed Executive Order NS2012 authorizing the Secretary of Agriculture to oversee the nation's supply of hemp for national security. (National Defense Resources Preparedness, § 201(a)(1).) The modern federal definition of industrial hemp comes from the legal definition of banned marijuana and the exemptions in the 2015 Farm Bill. That comes with some caveats, though. It only applies in states with industrial hemp laws and should be connected to university research.

The simpler definition is that if you use cannabis to make products for a psychotropic or medicinal effect, it is marijuana. If you use it for anything else — food, clothing, shelter, paper, plastic, fuel, agronomic value, etc. — it is hemp.

Credit: Aleks

Public Law No: 113-79. § 7606 Definitions (2) Industrial Hemp — The term "industrial hemp" means the plant Cannabis sativa L. and any part of such plant, whether growing or not, with a delta-9 tetrahydrocannabinol concentration of not more than 0.3 percent on a dry weight basis.

Growing hemp is good for our planet. It can be used to remediate contaminants from the soil. It anchors soil against erosion during times of heavy rain and flooding. It shades the soil to cool the earth. It uses less chemical fertilizers and pesticides. It actually chokes out weeds when they overtake a field. It brings soil nutrients closer to the surface. It adds mulch to the topsoil. It creates pathways into the subsoil that help forests regrow. It uses less water in times of drought. It provides food and nesting materials for birds and wildlife. Its fragrance sweetens the air. It sucks CO2 out of the atmosphere and releases oxygen.

There is growing evidence that it can be used to slow or reverse global warming, block ultraviolet radiation, and collect and absorb radioactive materials out of the environment. The economic benefits of living on a healthier planet might not fit into a spreadsheet, but they are considerable.

All these beneficial characteristics suggest that we should be following the instructions in George Washington's 1794 note for his Mt. Vernon

Industrial hemp grows a dense canopy that suppresses competing weeds. Credit: Kat_geb

gardener: "Make the most you can of the Indian Hemp seed and sow it everywhere." However, the global agronomic benefits are just a starting point for reasons that we should be growing as much hemp as possible. The agricultural reason is to provide sustainable, renewable natural products to meet humanity's needs for the necessities of life and other consumer goods.

A key issue facing the development of hemp industries is that they must currently rely almost exclusively on imported raw material. Hemp agriculture is in its domestic infancy, with just a few projects up and going in Kentucky and Colorado. That creates a financial burden, hamstrings U.S. agricultural interests and gives the DEA and customs departments a lot of latitude to mess with people's shipments into the country.

**EVERY INDUSTRY NEWBIE NEEDS TO KNOW:**

"If you set off to create a new hemp business, I suggest a careful look at the few successes and many failures among the 10,000 entrepreneurs who preceded you. I have seen hundreds of dedicated people open brick and mortar retail stores; almost every one has closed. Most who failed had no previous retail experience. Trying to distribute a product produced on another continent involves language barriers, letters of credit, currency differences, shipping and Customs problems, and a long, difficult supply chain. It will get better once we are able to produce hemp products here in the USA. Think of industrial hemp as a long, slow climb toward customer acceptance and legitimization." — *Don E. Wirtshafter*

## Small Businesses, Small Farmers vs. Big Government

In 2002 the DEA tried to ban hemp food products by changing the definition of marijuana set by Congress. After losing the HIA v DEA case in federal court, the Bush administration abandoned the plan to criminalize food products in 2004.

In May of 2014, the DEA seized 250 pounds of hemp seed imported from Italy for a special pilot hemp program legalized by the state of Kentucky. Agriculture Secretary James Comer asked Republican senators Rand Paul and Mitch McConnell to intervene on behalf of the state's farmers, and the DEA backed off. Kentucky's hemp program is still in operation.

## AGRICULTURE, HORTICULTURE

Hemp stalks are traditionally stacked in the field to dry. Credit: Andrei Chebotarev

The best way to grow hemp depends mostly on the desired end product. Cultivation for seed is quite different than cultivation for fiber. The first involves short plants with many branches that grow heavy with seed, while the second involves tall plants with thin stalks and few branches. For seed, the plants are set apart from each other to allow the lateral branches that will fill with flowers, anticipating the rich bounty of their seeds.

When harvesting seed, the seeds mature at different rates within each plant and within each crop. The trick with seed harvesting is to find a point in time when as many seeds as possible will be ripe and a minimum have split from their calyx pods and dropped to

Credit: Adrian Cable

## Shipping Basics

As when ordering an item, you want to get the net weight of your purchase, but also the shipping weight in a container. The difference between those two amounts is the tare weight. Whenever you are given a gross weight, be sure to find out the equivalent net rate or you might find yourself being shorted by a supplier. If you are a U.S. company doing business in the international community, you have to know the ratios between pounds and kilos, yards and meters, quarts and liters. It would be a sweet irony if the rise of the cannabis industry were to become the catalyst to bring the metric system to the U.S.

the ground. For fiber, the plants are sown or drilled densely in close rows that force the stalks to compete in a race for the sun's light. The trick with fiber harvesting is to keep the equipment running and the plant stalks parallel. When equipment jams or stalks become entangled, revenue losses ensue.

Shipping is a big part of the cost of doing agricultural business, so boost revenue by adding as much value as possible on-site. Due to processing and marketing considerations, seed oil products are easier to bring to market, with fiber crops still facing technical and infrastructural challenges; this has been the state of the U.S. hemp industry since the dismantling of the manufacturing base led to a loss of both factory equipment and skilled labor.

Hemp seed products tend to track in two directions: cold pressed vegetable oil and refined seed oil.

- Cold pressed oil is edible and intended for human consumption, but it has a short shelf life. The more clear, golden and lighter green varieties tend to be more filtered; darker green varieties tend to be more flavorful.
- Refined oil is golden to clear with a longer shelf life and sometimes subject to hydrogenation, heat and pressure during processing that make it less desirable for human ingestion but easier to mix with other compounds and still beneficial to the skin and hair as soap, shampoo or lip balm. Industrial-grade oil is intended for non-human use in lubricants, paints, sealants, inks, fuels, treatments for leather or wood, etc.

**EVERY INDUSTRY NEWBIE**
**NEEDS TO KNOW:**

Reduce shipping and processing costs while increasing your bargaining power by joining with your fellow hemp farmers in an agricultural cooperative.

## HEMP SEED PRODUCTS

The hemp seed products are the quickest to bring to market because of its relatively low mass to high value, and the fact that Canada has been producing and marketing seed products at a much faster rate than fiber products.

## FOOD RECIPES & MENUS

Including hemp in a food product is a statement of your personal or corporate commitment to making both the consumer and the earth more healthy and balanced. Not to be confused with cannabis edibles, hemp foods derive from the seeds and thus have no psychoactive effect. They won't make you fail a drug test, especially if they have the TestPledge seal.

Hemp seed can be consumed in its whole seed, hulled seed or one of its processed forms. Most people prefer the hulled seed; the whole seed shell has a gritty texture and a tendency to get stuck in the teeth. Still, there are enough minerals and fiber in the shell, including high-end dietary fibers, that whole seed deserves consideration.

The fruit of the cannabis plant has an exceptional nutritional content in its heart, with eight essential amino acids (proteins) and both omega 3 and omega 6 essential fatty acids (EFAs), as well as edestin, magnesium, zinc, iron and dietary fiber. Most commercial brands of hulled hemp seed have about 5 grams of protein in one tablespoon of hemp seeds — a higher content than either peanuts or almonds. Even so, it's their EFA content that make hemp seeds such nutritional champs. "Hemp seed oil may be nature's most perfectly balanced oil. It contains an ideal 3:1 ratio of omega-6s (linoleic acid) to omega-3s (alpha-linolenic acid) for long-term use, and provides the omega-6 derivative gamma-linolenic acid (GLA)," wrote Dr. Udo Erasmus in *Fats that Heal — Fats that Kill.*

Marketing a successful hemp food product can be as simple as starting with an existing commercial food product and then bringing hemp into the recipe. Sometimes it is as the hulled hemp seed, but most of the time you want to start with a more processed form of the seed to incorporate into your hemp food item, such as substituting hemp flour, seed oil or milk in place of the wheat or soy versions of the same ingredients.

You will need to identify the sources of hemp flour, oil, milk and butter, and make sure there is enough of a supply secured to support your product rollout plan. Due to the obvious public health issues, making

**EVERY INDUSTRY NEWBIE NEEDS TO KNOW:**

The ease of substituting hemp-derived products for the many ingredients already used in processed foods also opens up another ancillary niche in the hemp market: processing raw hemp seeds into the flours, oils and cakes incorporated by other companies into their recipes.

Items from the Hemp Hut travelling exhibit.

any food products is more tedious than many other products that are not intended for ingestion. This means complying with health code requirements, securing access to adequate commercial kitchen facilities and being subject to inspections that may result in fines or even a cease and desist order. It's the same as for any other food.

The fruit itself has a flavor that ranges from subtle to bland, which means that hemp seed is a great textural and nutritional supplement to practically any recipe — and especially in the health food and sports energy foods industries. Supplementing protein and healthy fats can be as easy as sprinkling hemp seed hearts on raw food products like salads. Other, more potent flavors can easily dominate hemp seed, when that is what you wish to achieve, or use your spices to draw upon and accentuate or emphasize the subtle nutty flavor of the seeds.

## Hemp Seed Oil Preparation

Whole seed stores for a long time in its protected shell. Hulled seed or oil intended for human consumption should be put through a no-heat process, stored in cold darkness and consumed within a few months. The temperature and storage of the seed are also important. Cooking with the oil destroys a lot of hemp seed oil's nutritional content so it's best to use cold, as for salad dressing.

## TOILETRIES & COSMETICS

Credit: Andrei Chebotarev

Toiletries and cosmetics can also be made from hemp seed. These include a variety of personal hygiene products like shampoos and skin creams that utilize the plant's rich lathering oils, which blend well with other plant oils and can be used as an ingredient to make assorted hygiene and personal care products like Dr. Bronner's soap, shampoos and conditioners, moisturizing creams, soothing lip balm, etc. Various ingredients like essential oils and flavors are added to enhance the user's experience, but because the seeds are typically 99.99 percent clean before being pressed into oil and most of these types of products are made with refined oil, any cannabis terpenes would have to be added back in to the mix. Since cannabis oil is high quality, fragile and may become rancid, many hemp seed oil products should be labeled with a use-by date.

## HEMP STALK APPLICATIONS

Hemp's sturdy stalk produces two kinds of fiber with value in commercial markets. The stalk's bark or bast fiber is very long and can be combed and spun into lightweight, high-tensile yarn or thread. The stalk's hurd, or cortex, is about 40 percent cellulose and has a woody core. The material is highly absorbent. Among the first hemp stalk products of the modern era were Stoned Wear shirts from China, hemp twine bracelets, and jewelry and animal bedding by the Dutch company HempFlax.

Bast fiber. Credit: Rasbak

## CLOTHING, CORDAGE & TEXTILES

For millennia, hemp fiber powered the world economy. Every town had a hemp walk to make rope. Its thread, rope and fabric went into maritime use for the sails, rigging and caulking of great ocean-going vessels as well as tents, clothing, windmills, nets, covered wagons, etc.

Traditional hemp outfit from China.

The fiber has numerous practical advantages, such as its great length, tensile strength, durability, traction, UV protection and biodegradability. It breathes, and absorbs and wicks away moisture.

Pure hemp fabric has a texture that is most similar to flax, and it has been used to make linens for many thousands of years. Hemp is most often used in blends that give it a huge range of appearances and uses, such as hemp-flax blends for lace, hemp-cotton blends for knits, and hemp-silk blends for lingerie and nightwear. It can be used unwoven for felts, in knits for stretch and insulation, woven cloth for smoothness, knitted for softness, or netted to incorporate open spaces for strength or design purposes. Hemp and cotton fibers are naturally of different lengths, with hemp fiber being the longer. Hemp fibers may be spun on equipment designed for flax to

> **EVERY NEWBIE NEEDS TO KNOW:**
>
> As with food, hemp textile products confer a reputation for caring about the environment and future of the planet because it can be grown with fewer pesticides, less fertilizers and less water than most other crops. As a producer or consumer, you need assurances that those values were practiced and not just assumed.

create beautiful 100 percent hemp linens but the equipment designed for spinning cotton requires short fibers, so therein lies a challenge. The response has often been to cottonize hemp, which allows it to be spun more easily with cotton but loses some of the cannabis plant's own characteristics of strength and endurance.

> An enduring problem for hemp first arose when Catherine Greene and Eli Whitney invented the cotton gin in the 1790s. Hemp and cotton fibers twist in opposite directions but the equipment is designed to favor cotton, so the modern response is often to cottonize hemp.

## A GROWING & SUSTAINABLE HOUSING MARKET

Of all the ways that hemp may save the world, the transformation of the housing market may be the most promising. In the skyrocketing housing market, as timber and other building materials become more scarce, hemp pressboards, insulating materials, composites and "hempcrete" offer the possibility of bringing the cost of building new housing back down to affordable levels.

Imagine the housing development of the near future. The contractor hires hemp farmers in agricultural areas near the construction site to create local jobs and save on transportation costs. The residents of the new community save on heating and cooling costs for the life of their residence — which is considerable, given their remarkable longevity. Lightweight hemp fiber pressboards have structural support capabilities that can surpass that of timber. Hemp composites for double pane window manufacture lose less heat than their metal equivalents.

But perhaps the material that has caused the most excitement is the so-called hempcrete formulation.

This grinds up the inner cortex or whole hemp stalk and mixes it with cement, lime and water to make a liquid concrete with a fraction of the weight that sand and gravel have — which vastly reduces transportation costs.

Hempcrete is not structurally weight-bearing, but it can be poured into molds and cast into panels or blocks to make walls, wielded with a trough, mixed with other materials or pigments, etc. The resulting material is little short of remarkable. Lightweight, durable, temperature and sound insulative, insect resistant, mildew resistant, breathability that reduces gas buildup inside an edifice, flexible to a degree ... an engineer, construction company or private contractor's dream building material. Its characteristics make hempcrete perfect for use in times with more extreme temperature shifts because it balances the internal temperature and is durable in event of natural disaster, broken up and remixed to rebuild and easily replaced as soon as large-scale hemp farming is underway. From the farmer's point of view, it uses the stalks without requiring decortication, retting (rarely done nowadays), combing, spinning, weaving and so forth as required for textiles.

The process even fixes CO2 as it sets, to further reduce the building's carbon footprint. The process was pioneered in France using a technology that was used in the building of the Roman Empire.

People are always going to need to live and work somewhere. You might want a house of your own someday, or if you have a house already you might like to save money on utilities, maintenance and repairs. As housing

Insulating boards made from hemp hurds. Credit: Suavegeot

costs rise, availability shrinks and old buildings fall into disrepair, domestic hemp production promises to rebuild our collapsing urban residential housing market by creating jobs and replacing blighted housing with energy efficient hemp residences.

At the same time, using hemp reduces the strain on and respects the true value of our forests, replaces and therefore frees up timber, concrete and other traditional building materials for use elsewhere in the infrastructure, like bridges, sewers and other places where they are needed. On a large enough scale, this offsets construction resources and farm-based market competition could make other raw materials more affordable.

## PAPER & 3D PRINTERS

The first paper invented by the Chinese and the Romans was made of hemp. The first book printed on a press, the *Gutenberg Bible*, was printed on hemp paper. Recycled hemp textiles became the "rag" paper of the Age of Enlightenment that was used to draft the Declaration of Independence. In 1916, the U.S. Department of Agriculture predicted that hemp paper could save America's forests and family farms. Today, hemp remains a niche market, its fiber and pulp being used for cigarette and specialty papers or added to recycled paper to increase its density and durability. But with the continued march of federal reform, hemp paper is poised for a comeback.

As a natural fiber, hemp can be bleached with hydrogen peroxide rather than the much more toxic chlorine. Although today's available hemp pulp is too limited in supply to be used on a large scale and is almost always mixed with other paper fibers, a domestic hemp supply could change that equation. The profit potential is certainly tempting; hemp pulp cost about six times more than tree pulp in 1994 and although the supply has increased since then, any producer today is just as likely to get high returns. The trick to making this work is doing as much of the processing as possible on the farm site — and quickly, so that most of the waste plant residue is returned to the soil and as little weight as necessary has to be shipped and stored.

Just as hemp is a good resource to make paper for our presses, printers and copiers, it is also a promising raw material for 3D printers to use in constructing items. The process works by creating forms by building up layers of "filament" material into a specific shape. The better quality of material in the printer, the better product it generates. Acrylonitrile Butadiene Styrene (ABS) plastic is the most common filament material used in 3D printing, followed by PLA (Polylactic Acid). Glass fiber is often used to increase the tensile strength of ABS plastic but, as with fiberglass, hemp fiber fulfills the same role more safely than does glass: greater flexibility, lighter weight, no sharp edges and less brittle. Hemp-based filament pellets or strands already are being fused together to construct or "print" the physical object. The high strength to weight ratio of hemp bast fiber helps keep product weight down.

**EVERY INDUSTRY NEWBIE NEEDS TO KNOW:**

The paper area is one that is relatively undeveloped and ripe for new technologies and ancillary services like portable pulping equipment that operates right in the field.

In both paper pulping and 3D filament production, the current state of affairs suggests that there is a lot of potential for advances in the near future.

## FUEL FOR CLEAN ENERGY DEMANDS

The energy market is another arena where hemp has great untapped potential. Biofuels, and especially corn-based ethanol, use plant carboyhdrates to replace fossil fuel hydrocarbons. Using the standard transesterification process, researchers at the University of Connecticut got a 97 percent efficient return to convert hemp seed oil into diesel fuel, and estimated that a large-scale hemp fiber farm would be able to fuel its own production using the seed the crop would produce.

Renewable hemp biomass has the potential to replace dirty coal-powered electricity. Credit: Arnold Paul

Hemp has plenty of more biofuel cards up its sleeve. Simple hemp biomass conversion processes can be as minimal as burning chopped stalks, pressing them into fuel pellets for appropriate burners, or converting the biomass into

*Pyrolysis* – n.

**High heat and pressure in the absence of oxygen to produce char, which is then used in place of coal or other charcoals.**

EVERY INDUSTRY NEWBIE
**NEEDS TO KNOW:**

Seed oil diesel fuel is significantly cleaner than petroleum versions and producing it on the farm would save transportation costs. The only "problem" is that hemp seeds are so valuable that there may not be any "excess" to convert. The financial incentive goes in the other direction.

charcoal through pyrolysis. Because it is not oxygenated, the resulting fuel burns readily. No fracking, no poisoned groundwater, no earthquakes — no brainer.

The conversion could be done at plants that produce charcoal from wood or coke from coal, but due to compaction and shipping costs it would be preferable to develop smaller systems that could be transported to the hemp fields and "fired up" there. Given the size of the energy market and demand, this is an area where an individual or a small local enterprise can provide their own hemp fuel on a relatively small scale — as opposed to the prevalent corn-derived ethanol model that requires massive capital investment, large-scale production and government subsidies. Hemp's options are more earth-friendly, people-friendly and wallet-friendly.

Hemp seed is also a good source of biodiesel, though health food market demand has made hempseed oil too valuable to use for fuel – for now.

Hemp's potential as a fuel crop shows promise but is still deeply in its research and development phase. Looking at the long term, however, once local supplies of hemp agricultural "waste" byproducts become available — meaning the male plants, discarded stalks, tangles of lower value fiber, hemicellulose pulp, what have you — begin to accumulate, someone in the right place with the right processing plan stands to have a huge supply of hemp biomass to convert into fuel. That person might be you.

The main ethical issue that comes into play with hemp power is that its conversion into fuel paradoxically requires fuels — like fossil fuels. The difference is that every year when the hemp crop is growing to feed, clothe, house and provide for humanity, the plants are all scrubbing CO2 out of the atmosphere and putting more oxygen right back into the air. If we can figure out how to fuel power cells and energy grids with locally produced and converted agricultural and community waste products, it will reduce our landfills, make the world more livable, and create jobs and profits in the process.

## COMPOSITES & FABRICATION

The benefits of hemp composites was first publicly demonstrated by that icon of business, Henry Ford. He built a car body out of hemp composites, and then smashed it with a sledge hammer in front of an amazed press corps to document the fact that it didn't even dent. Not only did Ford pay his employees well enough that they could afford to buy their own cars, he demonstrated that a car could be built profitably by American labor using resources that were grown on American soil. Times have changed but the guiding principle remains the same: Quality domestic products can create jobs and products that fuel consumption and economic growth.

The potential to make hemp composite materials is practically endless. Hempcrete combines hemp stalks with cement to make a building material. Mills use the broken stalk and fibers to make lighter weight yet stronger chipboards. The bast fibers combine with resin to make fiberboard that can be built up in layers or molded and pressed into forms — like fiberglass, except lighter weight, fewer dangerous edges when fractured and easier to dispose. These hemp composites are gaining ground in the manufacture of car component parts, especially for the interior of cars like Mercedes Benz and BMW to keep passengers comfortable normally and to better protect them in event of an accident. Look for hemp composites

Hemp composites make this glove compartment for a car. Credit: Christian Gahle

everywhere, from smoking pipes (of course) to briefcases for business people.

Hemp for everything — including kitchen sinks. Credit: Florian Gerlach

Plastics are a bit different from composites; they use synthetic or synthesized organics that can be poured and otherwise molded into any variety of waterproof shapes and forms. We see them all around us — fossil fuel versions of plastics have displaced many of our traditional resources, most notably hemp but also metal, glass, wood, leather and ceramics.

Plastics are so darn handy, they'd be really great if only they would go away once we throw them out. Fortunately, biodegradable hemp can replace nearly any plastic rope product if we should ever do anything crazy, like go back to using good, solid hemp ropes instead of slippery, stretchy nylon ropes. Another is to replace petrochemical hydrocarbons used to make plastics with organic carbohydrates. Whatever the job, hemp is up to the task.

Hemp harvester. Credit: Aleks

The lack of hemp-based feedstock is the perennial barrier, particularly in light of the availability of petroleum feedstock subsidized by massive tax breaks and loopholes to the oil industry. Visionaries like Henry Ford saw hemp biofuel and composite production concentrated in areas that already have high hemp production, so the supply of materials is secure. Eighty years later, Kentucky is out to retake the industry that it once controlled in the 19th century.

Colorado is testing the waters. California will be a powerhouse if it gets past its drought. Kansas and Nebraska have huge plans. If there is going to be a recovery of the American materials and manufacturing markets, there is a very good chance that being near the hemp fields will be a big advantage to any company that wants to source its product domestically.

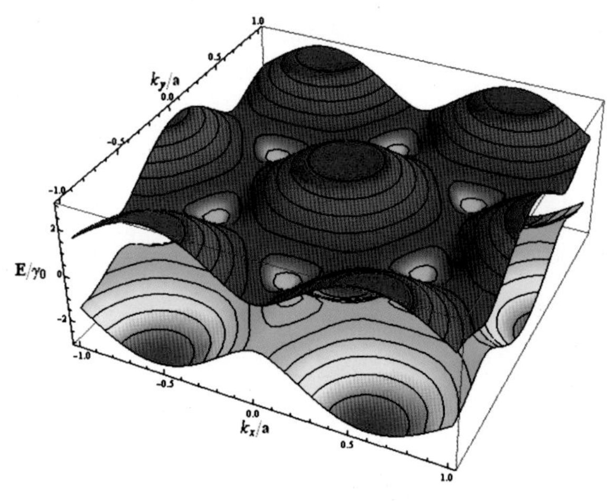


Hemp-derived graphene, a space-age material made of carbon atoms aligned in a single layer, could portend the future of hemp products. Credit: Paul Wenk

## Home-Grown Supercapacitors

A layer of carbon only one atom thick, stretched out and folded on itself thousands of times and capable of holding electrical charges powerful enough to drive a sports car — the future of batteries is nearly here. It's called graphene, and you can make it from hemp. When paired with abundant American supplies of sun and wind, the next generation of hemp-grown graphene super-batteries holds out the promise of a country fully self-dependent for its own energy: grown at home and built at home.

## CBD, ESSENTIAL OIL & TERPENE EXTRACTION

**Many people have looked at the possibility of extracting terpenes and cannabinoids —particularly CBD — from the industrial hemp crop. While we like the idea of taking advantage of hemp strains grown for other purposes to provide specific cannabinoid profiles, the idea of taking residual hemp from industrial hemp fields is probably not a viable strategy.** If you live on a hemp farm, you might pick up scraps and try to see what you can do with them on a micro-scale, but if you plan on growing strains for medicines, using medical grade bud and trim to make extracts is a more practical way to go. (See chapter on processing.)

The problem with using field hemp is that the cannabinoid content of the leaves and stalk are so low and there are also plenty of contaminants that might accumulate in the end product. For example, the plants that are collected throughout the season: were they sprayed? Did pesticides drift from other crops? The foliage, does it contain contaminants drawn up from the soil through the plant's vascular system? The equipment, was it of medical-grade purity or lubricated with petroleum oil? There are a lot of variables that need to be monitored, controlled and corrected, which isn't possible in every case.

Harvested hemp on a mesh fabric. Credit: Lossenelin

## EQUIPMENT, RESEARCH & DEVELOPMENT

Harvested stalks, colas attached. Credit: D-Kuru

Under the current legal regime, research and development are definitely the hot areas for the hemp industries. That is because the somewhat vague language may or may not allow much commercial production but it does make it abundantly clear that some university-sponsored research is possible. One bright point is that — whereas in Europe and Canada the hemp production on farms quickly exceeded the industrial capacity and had to be scaled back to allow industry to catch up — the lack of large-scale production gives the U.S. time for equipment and processes to be developed at a small scale and lay the groundwork to escalate production quickly as domestic demand grows.

It's easy to get lost in a mature field of fiber hemp. Credit: Barbetorte

Harvester cutting hemp stalks to gather seed.
Credit: Andrei Chebotarev

Breeding genomes that are drought resistant, flood resistant, weed and insect suppressive, etc., will be beneficial in the extreme weather that has become the norm of global climate change. Breeding strains that are especially good for soil remediation, absorb $CO_2$ from the atmosphere as a "carbon sink" or work as a fallow crop to improve and anchor soil all are critical, and if combined with phenotypes for easier bark separation, larger seed size and yield, higher cellulose content in the cortex and ease of harvest, numerous business sectors will benefit.

What areas are most ripe for development? First off, we need harvesting and processing equipment. Other countries have already converted grain planting and harvesting machinery for use with hemp, so researchers here can look at that data and make modifications before making any prototypes. Hemp has an extraordinarily tough bark (for an annual crop) and strong fiber (without qualification) that is difficult to cut, wears out blades quickly and gets tangled inside the equipment. To collect both seeds and stalk requires a harvesting machine that cuts and collects the tops separately from the rest of the stalks. We need faster processes to cut the hemp so as to make the bast fiber more valuable, and better equipment to decorticate and clean the fiber. We know that the cost of shipping the dried crop from the field to the factory greatly increases the cost for threshing, combing and spinning the bast fiber, and that having portable equipment for value-added processing at the farm site saves a lot of money in the back end. When the fiber crop is cut wrong or becomes entangled, it greatly increases the costs of processing the fiber and in some cases it essentially destroys its value. The woody core of the stalk is only about a half to a third usable for cellulose, and that the remaining plant binders and "hemicellulose" have very limited uses: return to the soil or use in hempcrete. What this tells us is that if you invent a system that makes it easy and affordable to separate, clean and

Congress made history in 2014 by allowing small research plots of hemp to be grown. The crimp in the plan is that it takes tens of thousands or hundreds of thousands of plants to do that type of genetic research, years to gauge its success and under the current law it is restricted to cultivars with less than 0.3 percent THC. Finding individual plants drawn from the high-THC strains that have a mutation or exhibit a recessive characteristic for low THC could be used to enhance the industrial hemp seed stock but there would always be the risk of a recessive high-THC gene showing up in the future. That is why the THC level restrictions need to be removed.

process the various parts of the plant, the market for that equipment or chemistry will increase as hemp production expands. If you come up with a product that capitalizes on the hemicellulose waste, there will be an abundant supply. All customers, regardless of size, will be looking for simplicity and affordability.

A popular media outlet has been providing timely information about cannabis for decades: Time4Hemp.com. That name captures the current status of the emerging industrial hemp markets. There are many challenges to bringing back this industry that helped build America, but one thing we know is that we did it before, during World War II (see: "How Did We Get Here?"), and we can do it again.

The challenge today — to find sustainable replacements for destructive and dwindling mining, forestry and fossil fuel resources before we wipe out life on earth — is greater than those facing any other generation, but we can do it.

## "Marihuana" Defined

USC Title 21 (Food and Drugs) Chapter 13 (Drug Abuse Prevention and Control) § 802: (16) The term "marihuana" means all parts of the plant Cannabis sativa L., whether growing or not; the seeds thereof; the resin extracted from any part of such plant; and every compound, manufacture, salt, derivative, mixture or preparation of such plant, its seeds or resin. Such term does not include the mature stalks of such plant, fiber provided from such stalks, oil or cake made from the seeds of such plant, any other compound, manufacture, salt, derivative, mixture or preparation of such mature stalks (except the resin extracted therefrom), fiber, oil or cake or the sterilized seed of such plant, which is incapable of germination.

Separating stalks in the Ukraine.

Harvesting fiber hemp.

Machine combing hemp fiber.
Credit: Larry Serbin

## Tolstoy on Hemp

Excerpt from Tolstoy's classic novel, *Anna Karenina:*

"Why are you up so early, my dear?" the old woman, their hostess, said, coming out of the hut and addressing him affectionately as an old friend. "Going shooting, granny. Do I go this way to the marsh?" "Straight out at the back; by our threshing floor, my dear, and hemp patches; there's a little footpath." Stepping carefully with her sunburnt, bare feet, the old woman conducted Levin, and moved back the fence for him by the threshing floor. "Straight on and you'll come to the marsh. Our lads drove the cattle there yesterday evening." ... The dew, not visible till the sun was up, wetted Levin's legs and his blouse above his belt in the high growing, fragrant hemp patch, from which the pollen had already fallen out. In the transparent stillness of morning the smallest sounds were audible. A bee flew by Levin's ear with the whizzing sound of a bullet. He looked carefully, and saw a second and a third. They were all flying from the beehives behind the hedge, and they disappeared over the hemp patch in the direction of the marsh.

# 17 LOOKING PAST PROHIBITION

Probably the toughest thing about writing a newbie's guide to cannabis and the industry is how fast everything is changing. Nonetheless, we are encouraged by major trends that have held steady for several years. Since the historic state legalization votes of 2012 and 2014, none of the prohibitionists' dire predictions have come true. Public safety is essentially unaffected or modestly improved. Badly-needed tax revenues have exceeded expectations. Public opinion continues to shift further in favor of more legalization.

The International Cannabis Business Conference in San Francisco. Credit: Alex Rogers

It's not a matter of *if* anymore, but *when* and *how*.

Americans seem to be reawakening to a passion that grips us every few generations or so, when we wake up to the plights of our fellow citizens who are not treated fairly by the law. People are aware that punishments for cannabis are undeserved, that people of color are selectively targeted under these policies and that some very sick patients are left to suffer just because the medicine they need fell victim to a smear campaign from 80 years ago. True to tradition, the people are demanding equality.

Professor Raphael Mechoulam.
Credit: Hebrew University

## THE CANNABIS CONSTITUENCY IS COMING OF AGE

There is a growing multi-partisan recognition that cannabis reform is a good way to win votes, especially with younger voters. So far it has been the Democrats that have provided the majority of reform legislation and votes nationally, with a handful of Republicans beginning to join in to give reform bills their margins of victory. Democrats Cory Booker, Kirsten Gillibrand and Gavin Newsom are cannabis advocates and rising stars of the party. However, some of the top leaders of the next wave of reform are starting to be Republicans. A CBS News poll found in November 2011 that only 27 percent of Republican voters thought marijuana should be legal but that 65 percent thought the issue should be left to state governments. By 2015, a Pew Research Center poll found that 63 percent of younger Republican "millennial" voters support legalization. GOP leaders who understand the vital importance of a youth issue that is tied to a locally grown, job-creating industry are beginning to draw the connections — especially if they happen to be running for president. Senators Rand Paul and Ted Cruz spoke up for states rights on the issue. Republican frontrunner Donald Trump said, "I think medical should happen—right? Don't we agree? I think so. And then I really believe we should leave it up to the states." Tough-talking New Jersey

**Governor Chris Christie promised a strong federal rollback of state laws and then slid precipitously in the polls.**

## A COSTLY WAR ON CANNABIS

On Capitol Hill, a narrative is gaining steam, even among people who hate marijuana the most, that sees some or most of the punitive federal measures as being far too burdensome to taxpayers. In short, there is a growing consensus in Congress that its multi-billion war on marijuana simply costs too much. One of the most decisive shifts has been the reframing of the "end of the drug war" argument as a conservative economic issue. It's just too expensive.

The costs of the drug war are not limited to the over $2 billion the DEA flushes down the nanny state toilet every year; there are the billions invested into prisons, billions spent by states, billions wasted by businesses on drug testing their employees, and millions of lives damaged or destroyed by this policy. There is also a massive economic loss in the U.S. of many billions of dollars that would be available by embracing one of the world's most rapidly expanding industries. ArcView Market Research's *The State of Legal Marijuana Markets* report points out that the U.S. market for legal cannabis grew 74 percent in 2014 to $2.7 billion (up from $1.5 billion in 2013), making it the fastest growing industry in the country. The report projects that if trends continue, when all states legalize marijuana the total market size should top $36.8 billion, making it larger than the organic food industry ($33.1 billion). Others predict that it could go three times that high. Industrial hemp would overshadow that several times over.

## LATE OUT OF THE GATE

Americans can learn from other countries. For the past 50 years, Israel has invested in medical marijuana research instead of drug wars. The Hebrew University laboratory led by Dr. Raphael Mechoulam has discovered much of what the world now knows about medical marijuana, including the structures of CBD, THC and anandamide. The investment has paid off; its veterans and citizens have access to cannabis for PTSD, nerve damage, cancer and other medical conditions. Not bad for a country with a population smaller than New York City. England's GW Pharmaceuticals is working to secure approval to market a cannabis plant-based medicine. Uruguay and

Jamaica are already moving ahead with progressive policies. Holland has been in the market for decades already.

So when the United States — with its vastly greater resources, workforce and world-class network of research universities — unties the hands of the American cannabis economy, there can be no doubt that we'll quickly take the lead. Legalization is coming; the potential to adapt to it is limitless and the opportunities are endless. Hopefully our leaders will have the foresight not to ratchet down on regulations that block access and ramp up fees and taxes to inflate prices. That will just perpetuate a criminalized underground economy to meet the expanding cannabis demand. Chart a better course.

## SPREAD THE WEALTH AROUND

It's not every day a government can so easily launch an entirely new multi-billion dollar industry from the ground up, to bring a nearly fully developed underground market out of the shadows and into the middle class. As we leave you here, right at the cusp of a new era in cannabis history, we just want to ask: what kind of foundation are we building for that future? We understand the interest in having strict regulations with robust oversight, but some of the policies are counterproductive and some of the resources used to enforce them would be better used going after serious crime and serious polluters.

It is our belief that more freedom is a good thing. Adults should be able to grow and consume a personal supply of cannabis and share with friends. That will undercut the financial incentives of the illicit market. Free enterprise with limited government oversight is part of our American heritage. The easiest approach would be legitimizing a broad sector of the nation's existing cottage industries of cannabis. If you're a public servant tasked with drafting regulations for a complex industry, be supportive and simplify the processes to bring the breeders, growers and distributors on board. Issue enough low volume licenses so people open shop and provide enough local access so the neighbors don't end up looking for their old dealers again. Issue enough high volume licenses to produce the competition that will bring prices down and frustrate the illicit market.

Cannabusinesses need to have practical regulation and operate with good, socially responsible and green business practices. Politicians need to court the cannabis constituency to earn their votes.

## THE FUTURE IS UP TO YOU

It should not need to be stated, but cannabis consumers deserve all their basic human rights to jobs, homes, health and family without facing bigotry and discrimination. We must do away with the penalties, drug testing and imposed stigma that allow society to violate the civil rights of cannabis consumers and providers. Cannabis consumers need to become politicized, call their elected officials, support candidates who serve their interests, perhaps run for office to guide the process. It's up to you.

Perhaps most of all, the cannabis consumers who are able to do so, need to come out of the closet and be ambassadors of change. More and more people who try cannabis are seen as responsible adults, and this social evolution appears to be going smoothly, except for all those people still being arrested or who are still sitting in prison. We have to do something about that. The harm done by the fraudulent drug war is deep and wide. Luckily, the cannabis plant has powers to heal that go even further.

If you're an entrepreneur, make us look good and feel good. Play by the rules —even if they chafe — but lobby for expansive reforms and support the groups, individuals and organizations that have done the legwork to carry us this far. Remember that this industry is built upon the backs of hundreds of thousands of people who have gone to prison and otherwise suffered for doing the very things that many of our readers aspire to do legally. The onus is on our cannabusinesses here, on the threshold of a new industry, to establish customs of mutual respect and uniform business standards. So provide high-quality goods and services at a fair price, be good to your community and good things will ensue. There's nothing wrong with national chains and brands, but remember we also need to give small businesses what they need so our neighborhoods will thrive.

If you're just trying a puff for the first time, take it easy. Relax. Slow down to savor your life and time. If that lesson were the only gift we ever got from cannabis, it would be enough; but instead it gives us all these things and more! It's finally time to give back, by dedicating your time, money and energy to advance reform and secure cannabis' rightful place in society.

Our thoughts go out to you, dear Newbie. May you flourish and surpass your expectations, and may cannabis be a benefit to you all your days.

Miko Sloper channels Uncle Sam's message. Photo by Chris Conrad.

# APPENDICES

**State vs. Federal Laws**

The laws of the United States are set up along a system of federalism, in which both state and federal laws exist simultaneously and parallel to one another. The U.S. Constitution sets up a relatively straightforward framework for dealing with any conflicts between the two: the Supremacy Clause provides for federal law to trump state law in case of a positive conflict, while the Ninth and Tenth Amendments limit federal powers to those specifically enumerated by the Constitution, with the remainder left to the states or the people. Yet as straightforward as this balance may be in concept, courts have often struggled to find the best way to balance competing state and federal interests in practice — with no issue posing thornier issues of federalism than the conflict between state and federal cannabis laws.

**The "Commerce Clause"**

Article II, Section 7 of the U.S. Constitution grants Congress the power to regulate commerce between states, thus articulating one of the "enumerated powers" mentioned above. So if there are goods in the flow of commerce that cross state or national lines, Congress may regulate their flow — so far, no problem.

But in the decision of *Gonzales v. Raich* (2005), the U.S. Supreme Court went further. In that case, a dispute between California medical marijuana patient Angel Raich and federal law enforcement over whether Raich could grow her own plants in her own yard for strictly her own use, the Court ruled 6-3 that Raich's garden could be regulated under the Commerce Clause

— despite the fact that Raich was not buying or selling anything. The six justices reasoned that although Raich engaged in no commercial activity to raise and consume her plants, the fact that she *might* have done so meant that, in the aggregate, she and all others growing cannabis for their own use in medical marijuana states could be "regulated" — which in this case is a euphemism for arrest, home invasion and the destruction of the only medicine that effectively treats her symptoms, thus condemning Raich to a lifetime of excruciating pain.

If this decision strikes you as giving Congress a blank constitutional check of unlimited power to do anything they want, well, you're not alone. The *Raich* decision has been sharply criticized by both drug policy reform advocates and constitutional watchdogs who have warned that the careful balance of powers articulated in the U.S. Constitution has been swept away by the decision, rendering the 9th and 10th Amendments effectively moot. Unfortunately, the only way to correct this atrocious reading is to either pass a constitutional amendment overruling it, or else insist that our future presidents only nominate federal judges who respect the constitution.

**The Power of the Purse**

But wait, it gets even more complicated. While federal cannabis law may be supreme over state reforms, and the president has a duty to execute the laws passed by Congress, the matter doesn't end there: all federal laws must be read in the context of the *funding* Congress provides the executive branch to carry them out.

For this reason, federal cannabis law has suffered from a case of multiple personality disorder for a long time. The first federal anti-cannabis law, the Marihuana Tax Stamp Act of 1937, was ostensibly a revenue law that at least theoretically paid for its own enforcement; but ever since the passage of the Controlled Substances Act of 1970, federal drug control has been a giant money hole that Congress either has — or occasionally, has not — voted to fund. From 1973, when Congress authorized $75 million to fund the Drug Enforcement Administration's operation, to 2015, when the bloated agency's funding exceeded $2 billion, Congress has continually heaped more and more tax dollars on an agency that has utterly failed to eliminate illegal drugs from America's streets and schools.

Finally, Congress has begun to exercise its constitutional duty and announced that enough is enough. In late 2014, President Barack Obama signed into law a spending bill that included an amendment which forbade the use of any funds by the Department of Justice to target state governments that had passed medical marijuana reform. The funding bill was renewed as this book went to press, but we hope that Congress will not only renew but expand the provision, so that Washington, D.C., can only use the taxpayers' money to mind their own business and not interfere either with the operation of state reforms or with the rights of individual patients. The framers of the Constitution might have been shocked to know that's even a debatable proposition.

**Executive Priorities**

Finally, given that the federal Justice Department has only those limited funds that Congress may grant for the funding of its activities, federal cops and prosecutors must exercise discretion in which cases to pursue. And here, the legal cannabis industry has some helpful advice from the office of the Attorney General.

On Aug. 29, 2013, Deputy Attorney General James Cole issued a memorandum to all U.S. Attorneys on which kinds of cannabis cases should be prioritized:

- Preventing the distribution of marijuana to minors;
- Preventing revenue from the sale of marijuana from going to criminal enterprises;
- Preventing the diversion of marijuana from states where it is legal under state law in some form to other states;
- Preventing state-authorized marijuana activity from being used as a cover or pretext for the trafficking of other illegal drugs or other illegal activity;
- Preventing violence and the use of firearms in the cultivation and distribution of marijuana;
- Preventing drugged driving and the exacerbation of other adverse public health consequences associated with marijuana use;
- Preventing the growing of marijuana on public lands and the attendant public safety and environmental dangers posed by marijuana production on public lands; and
- Preventing marijuana possession or use on federal property. (James M. Cole, "Guidance Regarding Marijuana Enforcement — Memorandum for all U.S. Attorneys," Aug. 29, 2013).

It is important to note that this guidance is advisory, not binding. In fact, Melinda Haag, U.S. Attorney for the Northern District of California, responded to the "Cole memo" by announcing that her prosecutions, including against Oakland's Harborside Health Center, which had violated none of the eight priorities, would go on. So beware rogue prosecutors: the only way to remove all federal risk from the equation is to change federal law.

**The FDA vs. the DEA**

There are many agencies involved in the "alphabet soup" of federal drug policy, but the three bureaucracies that cannabis entrepreneurs need to focus on most are the Food and Drug Administration (FDA), the Drug Enforcement Administration (DEA) and the Internal Revenue Service (IRS), with honorable mention to the National Institute on Drug Abuse (NIDA).

*The FDA: The Moat Around the Money Pile*

If you can afford it, believe us, you want to be an FDA-regulated cannabis business. The FDA has an extremely expensive regulatory model that has been described as "pay to play" by critics who point out that only the richest corporations can afford to satisfy the agency's stringent requirements for approval of new drugs (as of 2015, the average cost of introducing a new FDA-approved drug was about $350 million per drug, but when factoring in the cost that the vast majority of new drugs fail to meet the requirements, the cost of a successful approval averages out at closer to *$5 billion* per drug[1]). As of this writing, the number of companies attempting to develop cannabis medicines through the FDA process is very small, but we expect that number to grow soon.

The advantages to having an FDA-approved drug on the market are legion. The majority of doctors will only prescribe FDA-approved drugs, preferring to avoid the risk of recommending non-approved drugs (this is why an entirely separate industry of "pot docs" has cropped up in medical states like California). This means that any company that wants to target the majority of the massive U.S. healthcare market will need to go through the FDA.

FDA-approved drugs are also covered by insurance. As of this printing, no patient had yet paid for medical marijuana through the Affordable Care Act (Obamacare), but we have no doubt that is coming soon. This (admittedly inefficient) mechanism ensures that the corporations that develop the expensive FDA drugs can recoup their investment through insane margins that few patients would be able to pay for out of pocket. This allows British corporation GW Pharmaceuticals to sell Sativex, its patented cannabis extract and delivery device, at the princely sum of $16,000 per year overseas.[2]

Last but not least, FDA approval allows corporations to patent their drugs, which is another great way to prop up the price of your drug by eliminating competition from generic analogues — for a limited time at least. Contrary to popular belief, it is not actually legally possible to patent a plant (Monsanto conspiracy theorists take note), but it is possible to patent ways to cultivate or process a plant, or products that are made from a plant — like Sativex.

*The DEA: The Dinosaur with Teeth*

The Drug Enforcement Administration is, in one sense, the successor agency to the 1930s-era Federal Bureau of Narcotics (FBN) and the 1960s-era Bureau of Narcotics and Dangerous Drugs (BNDD). But in another sense, it's an entirely different animal.

Upon its creation in 1970, the DEA stepped into the same role as these other agencies — criminal enforcement of illegal drug laws — but made a sharp break from the government philosophy that originally established its predecessors. Whereas the first federal laws against cannabis and other prohibited drugs operated as tax policies overseen by the Department of the Treasury (see "How Did We Get Here?"), the passage of the Controlled Substances Act and the establishment of the DEA shifted federal cannabis enforcement from revenue agencies to strictly criminal justice bureaucracies like the FBI. Thus, cannabis entrepreneurs of today have to plan not only for tax compliance under the IRS (see below) but also provisions in the federal criminal code.

Unfortunately, as of this writing, there is no guaranteed way for cannabis entrepreneurs to get the DEA off their backs. Entrepreneurs can limit their risk of federal enforcement by following the latest Attorney General guidelines (see above) and by following best practices for community engagement (see "Stakeholders"). But free and clear operation unfortunately has to wait on federal regulations.

There is a ray of hope, however. As of press time, incoming DEA administrator Chuck Rosenberg has expressed an interest in downplaying marijuana enforcement at the agency in favor of focusing on more dangerous drugs.

*The IRS: The Once and Future Cop*

The Treasury Department has taken a central role in federal cannabis policy since 1937, and unfortunately for the cannabis industry, that doesn't appear likely to change any time soon. Hopefully, however, the nation's top tax collectors will soon eliminate the devastating double standard that has treated cannabis businesses like second-class citizens for almost two decades: Section 280E of the Internal Revenue Code.

Under the new interpretation of 280E, which was passed in the 1980s by Congress to persecute international drug cartels, "traffickers" in controlled substances are forbidden from making the same kind of standard business deductions that are available to any other kind of business under the U.S. tax code. That means that claiming rent, utilities, human resources and all kinds of other expenses will get you nowhere with the IRS; the federal taxman will levy you on your *gross* income, not *net* income. Unfortunately, this has resulted in some crushing bills from legal cannabis businesses that have been taxed practically into nonexistence.

*NIDA: Just Because*

The National Institute on Drug Abuse (NIDA) can be the thorniest bureaucracy in the side of any cannabis entrepreneur, but with any luck it should be going away soon.

Defying all logic, the federal government has determined that research into the properties of cannabis must be subject to an extra requirement that no other drug on earth has to pass through: NIDA approval. Whereas a researcher with a plan to study crack cocaine or heroin on human subjects only needs to apply for the proper IRBs and DEA licenses, anyone with a plan to study cannabis has no choice but go through every step necessary to study heroin *plus* an additional review by NIDA before the agency will agree to sell them cannabis from the only legal supply in the country. This has led to a curious situation in which Dr. Carl Hart, neuroscientist at Columbia University, can get permission to administer crack cocaine to his human subjects in his laboratory, while Dr. Sue Sisley of the University of Arizona and Rick Doblin, PhD, of the Multidisciplinary Association for Psychedelic Studies (MAPS) had to fight for years to get approval for their study on the effects of cannabis on combat veterans with treatment-resistant PTSD.

Fortunately, a key part of this bizarre and barbaric bureaucratic hurdle has just been removed. Days before press time, President Obama ordered an end to outdated and redundant Public Health Service (PHS) review of cannabis trials, so there's hope that he could do the same for the entire NIDA monopoly.

**The Schedules (and how to get off them)**

Fortunately, the CSA provides for a process by which cannabis and other drugs may be rescheduled. Unfortunately, in the case of cannabis, our government has repeatedly refused to exercise it — even when a federal judge told them to do so.

Congress set some of the drugs (including cannabis) into the CSA's schedules upon the law's passage, but then delegated further scheduling authority to the President via U.S. Attorney General, who in turn delegated that authority to the Administrator of the DEA. The CSA also provided that "any interested party" may petition the DEA to schedule, reschedule or deschedule any drug. This is a provision that cannabis advocates have used repeatedly, trying without avail to induce their government to uphold the terms of the original deal and reschedule cannabis as more information has become known about it.

*Schedule I*

To appreciate just how disingenuous this policy history is in context, let's examine what Schedule I means according to the terms of the CSA. Schedule I drugs supposedly have:

- No currently accepted medical value in the United States;
- An extremely high risk of abuse; and
- Such a dangerous risk of use that they cannot be used safely, even under a doctor's supervision.

This *may* be the place for drugs like heroin (which, a little known fact, was invented by the German pharmaceutical company Bayer). There are precious few scholars who still believe that these factors describe cannabis in the slightest. As of this writing, a bill that would reschedule cannabis to Schedule II is pending in the U.S. Senate.

Drugs in Schedule I (these lists are not exhaustive):

Heroin
LSD
Psilocybin (i.e., "magic mushrooms")
MDMA (under investigation as a PTSD treatment)
Mescaline (i.e., peyote or San Pedro cactus, subject to limited religious exemptions)
DMT (i.e., ayahuasca, subject to limited religious exemptions)

*Schedule II*

If the DEA's scheduling policy were driven by science and not politics, most drugs currently in Schedule I would actually be placed in Schedule II. These are far from the world's most awesome substances, but nevertheless they have at least some medical utility, at least in extreme cases like end-of-life compassionate care. Schedule II drugs have:

- A currently accepted medical use in treatment in the United States, or currently accepted medical use with severe restrictions;
- A high potential for abuse; and
- An abuse potential such that its use may lead to severe psychological or physical dependence.

Drugs in Schedule II:

Cocaine
PCP
Opium
Morphine
Barbiturates
"Street" amphetamines (i.e., "crystal meth")
Legal amphetamines (i.e., Adderall)

*Schedule III*

This schedule is for those drugs that could pose problems, and even serious problems, but don't quite rise to the threat level of the drugs in Schedule I/II. Ironically, pure tetrahydrocannabinol (THC), the principal psychoactive ingredient in most strains of cannabis, has been categorized in this schedule since 1999 (having spent the previous decade in Schedule II) when it comes in the form of Marinol, a petroleum-derived synthetic stuffed into a pill separate from all the other compounds that induce the plant's ensemble effect. So, according to federal law, the pure and unmitigated drug most responsible for cannabis' psychoactive effects (THC, Schedule III) is a legitimate medicine, but if it is foolishly combined with other cannabinoids (like CBD) that are *less* psychoactive and even go so far as to *reduce* the psychoactive effects of THC alone — well, somehow that turns an otherwise benign THC into a drug of abuse with no medical value (plant cannabis, Schedule I). Who knew?

Schedule III drugs have:

- A currently accepted medical use in treatment in the United States;
- A potential for abuse less than the drugs or other substances in Schedules I and II; and
- An abuse potential that may lead to moderate or low physical dependence or high psychological dependence.

Drugs in Schedule III:

Ketamine
Some kinds of barbiturates
Paregoric (opium mixed in camphor)
Marinol (synthetic THC)

*Schedule IV*

Down around Schedule IV, the listed drugs begin to look more and more harmless:

- A currently accepted medical use in treatment in the United States;
- A low potential for abuse relative to the drugs or other substances in Schedule III; and
- May lead to limited physical dependence or psychological dependence relative to the drugs or other substances in Schedule III.

Drugs in Schedule IV:

Benzodiazepenes like Valium, Xanax and Klonopin
Chloral hydrate, popular as a sedative in the 19th century

*Schedule V*

Drugs in the least restricted schedule are mostly obscure. They have:

- A currently accepted medical use in treatment in the United States;
- A low potential for abuse relative to the drugs or other substances in Schedule IV; and
- A limited risk of physical dependence or psychological dependence relative to the drugs or other substances in Schedule IV.

Drugs in Schedule V:

Cough suppressants containing codeine
Certain anti-diarrheal medications containing atropine

# GLOSSARY

**215:** shorthand for medical marijuana in deference to California's Prop 215 initiative.

**420:** means "it's time to get high" and originated in the 1970s from a meeting time for a group of students in San Rafael, California.

**710:** refers to oil; spelled upside down (flip the book over to see for yourself).

**Blunts:** joints rolled in tobacco or hollowed out cigars, with cannabis replacing some tobacco.

**Bud:** the mature flower inflorescence of a female cannabis plant. Green bud, kind bud and chronic are just a few of the words people use to refer to mature cannabis flowers.

**Budtenders:** point of purchase sales staff and consultants regarding the various kinds of cannabis and products offered at a particular location.

**Cannabinoids:** are specialized compounds found on the cannabis plant in acid form: metabolized THC has objective subjective and objective medical effects; CBD has objective medical effects; CBG is the precursor; CBN is the degraded form; CBC is also of some medical interest; others being studied.

**Clones:** plants formed by taking branch cuttings from a mother plant and rooting them.

**CO2:** carbon dioxide.

**Cola:** refers to the central stalk bud of cannabis.

**Decarboxylate:** means to remove a chemical chain from the plant molecule to metabolize it; most importantly for the reader, in the case of THC-acid, it must be decarboxylated to have a subjective psychotropic effect.

**Dioecious:** refers to plants with male and female flowers on separate individual plants.

**Dysphoria, dysphoric:** refers to the negative psychotropic effects of cannabis, such as causing paranoia.

**Euphoria, euphoric:** refers to the subjective positive psychotropic effects of cannabis, such as making you high.

**Flowers:** Cannabis plants are sexually distinct and can be staminate (male), pistillate (female) or hermaphrodite (both male and female), but only the females are usable for marijuana.

**Full Extract Cannabis Oil (FECO), Phoenix tears, Rick Simpson Oil (RSO):** Concentrated oil intended to be taken in large or small oral doses or to be applied topically.

**Genotype and Phenotype:** A plant's genotype refers to its underlying genetic makeup; its phenotype relates to its physical characteristics and traits.

**Hemp:** the proper English name for the cannabis plant, also distinguished as "true hemp."

**HID:** refers to high-intensity discharge lamps, usually either metal halide (MH) or high pressure sodium (HPS).

**IND, Investigational New Drugs:** program under which the federal government provides cannabis for medical use.

**Inflorescence:** the clustering of flowers along a plant's upper stalk.

**Joints:** cannabis cigarettes, sometimes with filters. Most commercial joints have filters.

**NPK ratio:** the mineral ratio in a fertilizer, essential to the plant's growth. N is for Nitrogen, P for Phosphorus and K for Potassium.

**Oaksterdam:** the cannabis-friendly area of downtown Oakland, California.

**Pharma, pharmaceutical:** refers to regulatory control over production and dispensing of all medical drugs that now consist of synthetic single molecule or compounded substances (typically administered via drops, needles or pills).

**Psychoactive:** refers to any substance that interacts with human brain chemistry, such as CBD.

**Psychotropic:** refers to a substance that affects conscious perception, such as THC, causing a subjective effect known as getting high.

**Seed line, cultivars:** refers to plant genotypes, also called varieties or strains.

**Seedlings:** young plants propagated from seeds that were harvested from parent plants.

**Sinsemilla:** Spanish for "without seed," it refers to a seedless, mature female cannabis bud. Unlike seedless watermelons and grapes, sinsemilla cannabis is not a genetic fluke; it is the result of withholding pollen from the flowering female plant.

**Resin:** refers to trichomes removed from the cannabis plant in various forms, including kief, hash, charas, concentrate or extracts.

**Terpenes:** volatile plant compounds that produce fragrance or taste.

**Toke, toking:** terms used to distinguish cannabis smoking from smoking tobacco.

**Total Dissolved Solids (TDS):** Refers to liquid nutrients and the minerals in your local hard water supply. You need to know this ratio to make sure you don't overfertilize your garden.

**Trichomes:** refers to the resin glands on a cannabis plant; the ones with round heads on them are called capitates.

**Vaporize, vape:** to heat cannabis or resin to a temperature where it decarboxylates and can be inhaled without burning so that one receives the medical or psychotropic effect without emitting the ash, smoke and particulate caused by smoking.

**Weed:** one of many references to cannabis. There are hundreds or even thousands of code words and pet names for cannabis.

# RECOMMENDED RESOURCES

**Business Events**

Cannabis Business World Congress, wcbexpo.com; B2B plus politics and investors in New York and Los Angeles every year.

CannaGrow Expo, cannagrowexpo.com; the art and science of cultivation.

CannaConnections.Events, site listing for events.

CHAMPS Trade Shows, champstradeshows.com; counterculture B2B expo.

IndoExpo Trade Show & Conference, indoexpoco.com; for growers, in Denver twice a year.

International Cannabis Business Conference, internationalcbc.com

Marijuana Business Expo, mjbizconference.com; B2B plus investors, in Chicago and Las Vegas every year.

National Cannabis Industry Association, thecannabisindustry.org; hosts regular events.

Cannabis Business Summit and Expo, cannabisbusinesssummit.com

State of Marijuana Conference, facebook.com/stateofmj

Women Grow, womengrow.com; connect, educate, empower cannabis industry leaders.

**Cannabis Events, Cultural & Research**

Clinical Conference on Cannabis Therapeutics/Patients Out of Time, medicalcannabis.com

Emerald Cup, theemeraldcup.com; Emerald Triangle grow contest and expo.

High Times Cannabis Cup, medcancup.com; contest for best cannabis and expo.

HempCon, Hempcon.com

International Association for Cannabinoid Medicines, cannabis-med.org

International Cannabinoid Research Society, icrs.co

International Drug Policy Reform Conference, drugpolicy.org

Minority Cannabis Business Association, minoritycannabis.org

Seattle Hempfest, hempfest.org; largest "protestival" in the U.S. every August.

**Publications**

*Cannabis Now Magazine* (B2C), cannabisnowmagazine.com

*Culture* (B2C), ireadculture.com

*High Times* (B2C), hightimes.com

*Marijuana Business Magazine* (B2B), mjbizmagazine.com

*Marijuana Venture Magazine* (B2B), marijuanaventure.com

*Skunk Magazine*, skunkmagazine.com

**Websites**

Center for Medicinal Cannabis Research, cmcr.ucsd.edu

Chris Conrad, chrisconrad.com

Cannabis Culture, cannabisculture.com

Celebstoner.com

Drug War Chronicles, stopthedrugwar.org

Drug War Facts, drugwarfacts.org

Green Flower Media, greenflowermedia.com

Ladybud Magazine, ladybud.com

Leafly.com; search for dispensaries, doctors, strains, products, news.

MarijuanaMajority.com

TheLeafOnline.com; news and analysis of evolving policies.

Weedmaps.com; search for dispensaries and doctors.

VeryImportantPotheads.com

**Organizations**

American Herbal Products Association (AHPA), ahpa.org

Americans for Safe Access (ASA), safeaccessnow.org
California NORML, canorml.org
Cannabis Consumers Campaign, cannabisconsumers.org
Drug Policy Alliance (DPA), drugpolicy.org
Family Council on Drug Awareness, fcda.org
Hemp Industries Association (HIA), thehia.org
Marijuana Policy Project (MPP), mpp.org
Multidisciplinary Association of Psychedelic Studies (MAPS), maps.org
National Cannabis Industry Association (NCIA), thecannabisindustry.org
National Organization for the Reform of Marijuana Laws (NORML), norml.org
Society of Cannabis Clinicians, cannabisclinicians.org
Students for Sensible Drug Policy (SSDP), ssdp.org
Vote Hemp, votehemp.com

**Radio, Podcasts**

Drug Truth Network/Cultural Baggage, drugtruth.net
Leaf Radio, timeforhemp.com
The Russ Belville Show, radicalruss.com
TimeforHemp.com; network of various radio show and podcast hosts.

**Miscellaneous**

Oaksterdam University, oaksterdamuniversity.com
THC Staffing Group, thcstaffinggroup.com; employment services in the cannabis industry.
WeedHire.com; job opportunities.
Weedgear.com; products for cultivation and consumption, newsfeed.

**Books for Further Reading:**

*American Herbal Pharmacopoeia: Cannabis Inflorescence*, 2014.
*Aunt Sandy's Medical Marijuana Cookbook: Comfort Food for Mind and Body*, by Sandy Moriarty, Quick American, 2010.
*Beyond Buds: Marijuana Extracts – Hash, Vaping, Dabbing, Edibles & Medicines*, by Ed Rosenthal with David Downs, Quick American, 2014.
*Cannabis Pharmacy: The Practical Guide to Medical Marijuana*, by Michael Backes, Black Dog and Leventhal, 2014.
*Cannabis Yields and Dosage*, by Chris Conrad, Creative Xpressions, 2015.
*Hemp Bound: Dispatches from the Front Lines of the Next Agricultural Revolution*, by Doug Fine, Chelsea Green Publishing, 2014.
*Hemp for Health*, by Chris Conrad, Inner Traditions, 1997.

*Hemp: Lifeline to the Future*, by Chris Conrad, Creative Xpressions, 1994 (to be updated and re-released).

*Marijuana: Gateway to Health*, by Clint Werner, Dachstar Press, 2011.

*Marijuana Grower's Handbook*, by Ed Rosenthal, Quick American, 2010.

*Marijuana Horticulture: The Indoor/Outdoor Medical Grower's Bible*, by Jorge Cervantes, Van Patten Publishing, 2007.

*Marijuana is Safer: So why are we driving people to drink*, by Paul Armentano, Mason Tvert, Steve Fox, Chelsea Green Publishing, 2009.

*Marijuana Medical Handbook: Practical Guide to Therapeutic Uses of Marijuana*, by Dale Gieringer, Ph.D., Ed Rosenthal, and Gregory T. Carter, M.D., Quick American Publishing 2008.

*Mary Jane: The Complete Marijuana Handbook for Women*, by Cheri Sicard, Seal Press, 2015.

*Smoke Signals: A Social History of Marijuana - Medical, Recreational and Scientific*, by Martin Lee, Scribner, 2012.

*The Cannabis Encyclopedia: the definitive guide to cultivation & consumption of medical marijuana*, by Jorge Cervantes, Van Patten Publishing, 2015.

*The Cannabis Gourmet Cookbook*, by Cheri Sicard, Z Dog Media, 2012.

*The Emperor Wears No Clothes, The Authoritative Historical Record of Cannabis and the Conspiracy Against Marijuana*, by Jack Herer.

*The Marihuana Conviction*, by Richard Bonnie and Charles Whitebread II, The Lindesmith Center, 1999.

*The Pot Book: A Complete Guide to Cannabis*, by Julie Holland, M.D., Ed. Park Street Press, 2010.

*Too High to Fail: Cannabis and the New Green Economic Revolution*, by Doug Fine, Gotham, 2013.

*Understanding Marijuana: A new look at the scientific evidence*, by Mitch Earleywine, Oxford Press, 2002.

*Weed the People: From Founding Fiber to Forbidden Fruit*, by Jeremy Daw, Cannabis Now Publications, 2012.

# DEDICATION

*To the prisoners of the drug war, to those who have lived lives of deprivation and suffering due to cannabis prohibition, to those who have fought the good fight to restore cannabis to its rightful place in society, and to those who have passed away without seeing the final victory.*

# ACKNOWLEDGEMENTS

Acknowledgements, thanks and appreciation go out to: Donald Abrams, Michael and Michelle Aldrich, Paul Armentano, ASA, David Bearman, Dean Becker, David Borden, David Bronner, Sandee Burbank, Karen Byars, Jorge Cervantes, Mitchell Colbert, DPA, Richard Davis, Ellen Douglas, Lyster Dewey, Ben and Alan Dronkers, Don Duncan, John Thomas Ellis, Marc and Jodie Emery, Robert Field, David R. Ford, Jeff and Dale Jones, Gatewood Galbraith, Dale Gieringer, Zenia Gilg, Debbie Goldsberry, Mark Greer, Lester Grinspoon, Harborside Health Center, Hemp Industries Association, Hempfest organizers and volunteers, Hempsters everywhere, Jack Herer, *High Times* magazine, Ellen Holland, David Hua, the Internet, Mary Pat and Monty Jacobs, Rob Kampia, Brenda Kershenbaum, Ellen Komp, Michael Krawitz, Kyle Kushman, Casper Leitch, Law Enforcement Against Prohibition, Richard Lee, Frank Lucido, Bill Maher, Gracie Malley, Jennifer Martin, Joe McNamara, Doug McVay, Raphael Mechoulam, The Medicine Man, Tod Mikuriya, MPP, Elvy Musikka, Ethan Nadelmann, NCIA, NORML, Oaksterdam University students and staff, Carl Olsen, Judy and Lynn Osburn, William Panzer, Caroljo Papac, Patients Out of Time, Amy Povah, Virginia Resner, Alex Rogers, Marsha Rosenbaum, Irv Rosenfeld, Ed Rosenthal, Ethan Russo, Aseem Sappal, Danielle Schumacher, Larry Serbin, SSDP, Cliff Schaffer, John Shafer, Steph Sherer, April Short, Alex Shum and Stoned Wear, Phil Smith, Sid Solomon, Paul Stanford, Eric Steenstra, Eric Sterling, Jeffrey Stonehill, Keith Stroup, Barry Stull, theLeafOnline.com, Andrea and Maria Tischler, John Vasconcellos, VoteHemp, Kirk Warren, Clint Werner, Lennice Werth, Jackie G. Wilson, Don Wirtshafter, Kevin Zeese, George Zimmer, our families, friends, colleagues and all who have given us information and inspiration over the years. Special thanks to Chris' beloved wife Mikki Norris, whose many years of support and keen eye in proofreading this book have been so useful (sativa) to our process.